DOMAIN-THEORETIC FOUNDATIONS
OF FUNCTIONAL PROGRAMMING

DOMAIN-THEORETIC FOUNDATIONS
OF FUNCTIONAL PROGRAMMING

Thomas Streicher

Technical University Darmstadt, Germany

 **World Scientific**

NEW JERSEY · LONDON · SINGAPORE · BEIJING · SHANGHAI · HONG KONG · TAIPEI · CHENNAI

Published by

World Scientific Publishing Co. Pte. Ltd.
5 Toh Tuck Link, Singapore 596224
USA office: 27 Warren Street, Suite 401-402, Hackensack, NJ 07601
UK office: 57 Shelton Street, Covent Garden, London WC2H 9HE

British Library Cataloguing-in-Publication Data
A catalogue record for this book is available from the British Library.

ISBN-13 978-981-270-142-8
ISBN-10 981-270-142-7

Printed in Singapore

dedicated to Dana Scott and Gordon Plotkin
who invented domain theory and logical relations

Contents

Preface

This little book is the outcome of a course I have given over the last ten years at the Technical University Darmstadt for students of Mathematics and Computer Science. The aim of this course is to provide a solid basis for students who want to write their Masters Thesis in the field of Denotational Semantics or want to start a PhD in this field. For the latter purpose it has been used successfully also at the Univ. of Birmingham (UK) by the students of Martin Escardó.

Thus I think this booklet serves well the purpose of filling the gap between introductory textbooks like e.g. [Winskel 1993] and the many research articles in the area of Denotational Semantics. Intentionally I have concentrated on denotational semantics based on *Domain Theory* and neglected the more recent and flourishing field of *Game Semantics* (see [Hyland and Ong 2000; Abramsky et.al. 2000]) which in a sense is located in between Operational and Denotational Semantics. The reason for this choice is that on the one hand Game Semantics is covered well in [McCusker 1998] and on the other hand I find domain based semantics mathematically simpler than competing approaches since its nature is more abstract and less combinatorial. Certainly this preference is somewhat subjective but my excuse is that I think one should write books rather about subjects which one knows quite well than about subjects with which one is less familiar.

We develop our subject by studying the properties of the well known functional kernel language PCF introduced by D. Scott in the late 1960ies. The scene is set in Chapters 2 and 3 where we introduce the operational and domain semantics of PCF, respectively. Subsequently we concentrate on studying the relation between operational and domain semantics employing more and more refined *logical relation* techniques culminating in the construction of the fully abstract model for PCF in Chapters 11 and

12. I think that our construction of the fully abstract model is more elegant and more concise than the accounts which can be found in the literature though, of course, it is heavily based on them. Somewhat off this main thread we show also how to interpret recursive types (Chapter 9) and give a self contained account of computability in Scott domains (Chapter 13) where we prove the classical theorem of [Plotkin 1977] characterizing the computable elements of the Scott model of PCF as those elements definable in PCF extended by two parallel constructs por ("parallel or") and ∃ (Plotkin's "continuous existential quantifier") providing an extensional variant of the *dove tailing* technique known from basic recursion theory.

Besides basic techniques like naive set theory, induction and recursion (as covered e.g. by [Winskel 1993]) we assume knowledge of basic category theory (as covered by [Barr and Wells 1990] or the first chapters of [MacLane 1998]) from Chapter 9 onwards and knowledge of basic recursion theory only in the final Chapter 13. Except these few prerequisits this little book is essentially self contained. However, the pace of exposition is not very slow and most straightforward verifications—in particular at the beginning—are left to the reader. We recommend the reader to solve the many exercises indicated in the text whenever they show up. Most of them are straightforward and in case they are not we give some hints.

I want to express my gratitude to all the colleagues who over the years have helped me a lot by countless discussions, providing preprints etc. Obviously, this little book would have been impossible without the seminal work of Dana Scott and Gordon Plotkin. The many other researchers in the field of domain theoretic semantics who have helped me are too numerous to be listed here. I mention explicitly just Klaus Keimel and Martin Escardó, the former because he was and still is the soul of our little working group on domain theory in Darmstadt, the latter because his successful use of my course notes for his own teaching brought me to think that it might be worthwhile to publish them. Besides for many comments on the text I am grateful to Martin also for helping me a lot with TEXnical matters. I acknowledge the use of Paul Taylor's diagram and prooftree macros which were essential for type setting.

Finally I want to thank the staff of IC press for continuous aid and patience with me during the process of preparing this book. I have experienced collaboration with them as most delightful in all phases of the work.

Chapter 1

Introduction

Functional programming languages are essentially as old as the more well-known imperative programming languges like FORTRAN, PASCAL, C etc. The oldest functional programming language is LISP which was developed by John McCarthy in the 1950ies, i.e. essentially in parallel with FOR-TRAN. Whereas *imperative* or *state-oriented* languages like FORTRAN were developed mainly for the purpose of *numerical computation* the intended area of application for functional languages like LISP was (and still is) the algorithmic manipulation of *symbolic data* like lists, trees etc.

The basic constructs of imperative languages are commands which modify state (e.g. by an assignment $x{:=}E$) and conditional iteration of commands (typically by **while**-loops). Moreover, imperative languages strongly support *random access* data structures like arrays which are most important in numerical computation.

In *purely functional languages*, however, there is no notion of state or state-changing command. Their basic concepts are

- application of a function to an argument
- definition of functions either *explicitly* (e.g. $f(x) = x{*}x{+}1$) or *recursively* (e.g. $f(x) = $ **if** $x{=}0$ **then** 1 **else** $x{*}f(x{-}1)$ **fi**).

These examples show that besides application and definition of functions one needs also basic operations on basic data types (like natural numbers or booleans) and a conditional for definition by cases. Moreover, all common functional programming languages like LISP, Scheme, (S)ML, Haskell etc. provide the facility of defining *recursive data types* by explicitly listing their constructors as e.g. in the following definition of the data type of binary trees

$$\mathsf{tree} = \mathsf{empty}() \mid \mathsf{mk\_tree}(\mathsf{tree}, \mathsf{tree})$$

where empty is a 0–ary constructor for the empty tree with no sons and mk_tree is a binary constructor taking two trees t_1 and t_2 and building a new tree where the left and right sons of its root are t_1 and t_2, respectively. Thus functional languages support not only the recursive definition of functions but also the recursive definition of data types. The latter has to be considered as a great advantage compared to imperative languages like PASCAL where recursive data types have to be implemented via pointers which is known to be a delicate task and a source of subtle mistakes which are difficult to eliminate.

A typical approach to the development of imperative programs is to design a *flow chart* describing and visualising the *dynamic behaviour* of the program. Thus, when programming in an imperative language the main task is to organize *complex dynamic behaviours*, the so–called *control flow*.

In functional programming, however, the dynamic behaviour of programs need not be specified explicitly. Instead one just has to *define* the function to be implemented. Of course, in practice these function definitions are fairly hierarchical, i.e. are based on a whole cascade of previously defined auxiliary functions. Then a *program* (as opposed to a function definition) usually takes the form of an application $f(e_1, \ldots, e_n)$ which is *evaluated* by the interpreter[1]. As programming in a functional language essentially consists of defining functions (explicitly or recursively) one need not worry about the dynamical aspects of execution as this task is taken over completely by the interpreter. Thus, one may concentrate on the *what* and forget about the *how* when programming in a functional language. However, when defining functions in a functional programming language one has to stick to the *forms of definition* as provided by the language and cannot use ordinary set-theoretic language as in everyday mathematics.

In the course of these lectures we will investigate functional (kernel) languages according to the following three aspects

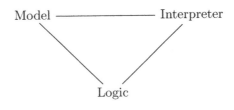

---

[1]But usually implementations of functional languages also provide the facility of compiling your programs.

or

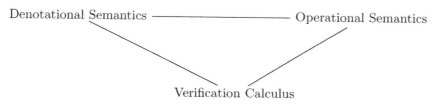

respectively and, in particular, how these aspects interact.

First we will introduce a most simple functional programming language PCF (Programming Computable Functionals) with natural numbers as base type but no general recursive types.

The *operational semantics* of PCF will be given by an *inductively* defined *evaluation relation*

$$E \Downarrow V$$

specifying which expressions E *evaluate* to which values V (where values are particular expressions which cannot be further evaluated). For example if $E \Downarrow V$ and E is a closed term of the type **nat** of natural numbers then V will be an expression of the form $\underline{n}$, i.e. a canonical expression for the natural number n (usually called *numeral*). It will turn out as a property of the evaluation relation $\Downarrow$ that $V_1 = V_2$ whenever $E \Downarrow V_1$ and $E \Downarrow V_2$. That means that $\Downarrow$ is *determinstic* in the sense that $\Downarrow$ assigns to a given expression E at most one value. An operational semantics as given by an (inductively defined) evaluation relation $\Downarrow$ is commonly called a "Big Step Semantics" as it abstracts from intermediary steps of the computation (of V from E).[2] Notice that in general there does not exists a value V with $E \Downarrow V$ for arbitrary expressions E, i.e. not every program terminates. This is due to the presence of *general recursion* in our language PCF guaranteeing that *all* computable functions on natural numbers can be expressed by PCF programs.

Based on the big step semantics for PCF as given by $\Downarrow$ we will introduce a notion of *observational equality* for closed PCF expressions of the same type where E_1 and E_2 are considerd as observationally equal iff for all contexts $C[\,]$ of base type **nat** it holds that

$$C[E_1] \Downarrow \underline{n} \quad \Longleftrightarrow \quad C[E_2] \Downarrow \underline{n}$$

---

[2]For sake of completeness we will also present a "Small Step Semantics" for PCF as well as an abstract machine serving as an interpreter for PCF.

for all natural numbers $n \in \mathbb{N}$. Intuitively, expressions E_1 and E_2 are observationally equal iff the same observations can be made for E_1 and E_2 where an *observation* of E consists of observing that $C[E]\Downarrow\underline{n}$ for some context $C[\,]$ of base type **nat** and some natural number n. This notion of observation is a mathematical formalisation of the common practice of *testing of programs* and the resulting view that programs are considered as (observationally) equal iff they pass the same tests.

However, this notion of observational equality is not very easy to use as it involves quantification over all contexts and these form a collection which is not so easy to grasp. Accordingly there arises the desire for more convenient criteria sufficient for observational equality which, in particular, avoid any reference to (the somewhat complex) syntactic notions of evaluation relation and context.

For this purpose we introduce a so-called *Denotational Semantics* for PCF which assigns to every closed expression E of type σ an element $[\![E]\!] \in D_\sigma$, called the *denotation* or *meaning* or *semantics* of E, where D_σ is a previously defined structured set (called "semantic domain") in which closed expressions of type σ will find their interpretation.

The idea of denotational semantics was introduced end of the 1960ies by Ch. Strachey and Dana S. Scott. Of course, there arises the question of what is the nature of the mathematical structure one should impose on semantical domains. Although the semantic domains which turn out as appropriate can be considered as particular topological spaces they are fairly different[3] in flavour from the spaces arising in analysis or geometry. An appropriate notion of semantic domain was introduced by Dana S. Scott who also developed their basic mathematical theory to quite some extent of sophistication. From the early 1970ies onwards various research groups all over the world invested quite some energy into developing the theory of semantic domains—from now on simply referred to as *Domain Theory*— both from a purely mathematical point of view and from the point of view of Computer Science as (at least one) important theory of meaning (semantics) for programming languages.

Though discussed later into much greater detail we now give a preliminary account of how the domains D_σ are constructed in which closed terms of type σ find their denotation. For the type **nat** of natural numbers one puts $D_{\mathbf{nat}} = \mathbb{N} \cup \{\bot\}$ where $\bot$ (called "bottom") stands for the denotation

---

[3]In particular, as we shall see they will not satisfy Hausdorff's separation property requiring that for distinct points x and y there are disjoint open sets U and V containing x and y, respectively.

of terms of type **nat** whose evaluation "diverges", i.e. does not terminate. We think of $D_{\mathbf{nat}}$ as endowed with an "information ordering" $\sqsubseteq$ w.r.t. which $\perp$ is the least element and all other elements are incomparable. The types of PCF are built up from the base type **nat** by the binary type forming operator $\to$ where $D_{\sigma \to \tau}$ is thought of as the type of (computable or continuous) functionals from D_{σ} to D_{τ}, i.e. $D_{\sigma \to \tau} \subseteq D_{\tau}^{D_{\sigma}} = \{f \mid f : D_{\sigma} \to D_{\tau}\}$. In particular, the domain $D_{\mathbf{nat} \to \mathbf{nat}}$ will consist of certain functions from $D_{\mathbf{nat}}$ to itself. It will turn out as appropriate to define $D_{\mathbf{nat} \to \mathbf{nat}}$ as consisting of those functions on $\mathbb{N} \cup \{\perp\}$ which are monotonic, i.e. preserve the information ordering $\sqsubseteq$. The clue of Domain Theory is that domains are not simply sets *but sets endowed with some additional structure* and $D_{\sigma \to \tau}$ will then accordingly consist of all *structure preserving* maps from D_{σ} to D_{τ}. However, for higher types (i.e. types of the form $\sigma \to \tau$ where σ is different form **nat**) it will turn out that it is not sufficient for maps in $D_{\sigma \to \tau}$ to preserve the information ordering $\sqsubseteq$. One has to require in addition some form of *continuity*[4] which can be expressed as the requirement that certain suprema are preserved by the functions. The information ordering on $D_{\sigma \to \tau}$ will be defined pointwise, i.e. $f \sqsubseteq g$ iff $f(x) \sqsubseteq g(x)$ for all $x \in D_{\sigma}$.

Denotational semantics provides a purely *extensional* view of functional programs as closed expressions of type $\sigma \to \tau$ will be interpreted as particular *functions* from D_{σ} to D_{τ} which are considered as equal when they deliver the same result for all arguments. In other words the meaning of such a program is fully determined by its input/output behaviour. Thus, denotational semantics just captures *what* is computed by a function (its extensional aspect) and abstracts from *how* the function is computed (its intensional aspect as e.g. time or space complexity).

When a programming language like PCF comes endowed with an operational and a denotational semantics there arises the question how good they fit together. We will now discuss a sequence of criteria for "goodness of fit" of increasing strength.

Correctness

Closed expressions P and Q of type σ are called *semantically* or *denotationally equal* iff $[\![P]\!] = [\![Q]\!] \in D_{\sigma}$. We call the operational semantics *correct* w.r.t. the denotational one iff P and V are denotationally equal whenever $P \Downarrow V$, i.e. when evaluation preserves semantical equality. In particular for

---

[4]which is in accordance with the usual topological notion of continuity when the domains D_{σ} and D_{τ} are endowed with the so-called *Scott topology* which is defined in terms of the information ordering

programs, i.e. closed expressions P of base type **nat**, correctness ensures that $[\![P]\!] = n$ whenever $P \Downarrow \underline{n}$, i.e. the operational semantics evaluates a program in case of termination to the number which is prescribed by the denotational semantics.

Completeness

On the other hand it is also desirable that if a program denotes n then the operational semantics evaluates program P to the numeral $\underline{n}$ or, more formally, $P \Downarrow \underline{n}$ whenever $[\![P]\!] = n$ in which case we call the operational semantics *complete* w.r.t. the denotational semantics.

Computational Adequacy

In case the operational semantics is both correct and complete w.r.t. the denotational semantics, i.e.

$$P \Downarrow \underline{n} \iff [\![P]\!] = n$$

for all programs P and natural numbers n, we say that the denotational semantics is *computationally adequate*[5] w.r.t. the operational semantics.

Computational adequacy is sort of a minimal requirement for the relation between operational and denotational semantics and holds for (almost) all examples considered in the literature. Nevertheless, we shall see later that the proof of computational adequacy does indeed require some mathematical sophistication.

If the denotational semantics is computationally adequate w.r.t. the operational semantics then closed expressions P and Q are observationally equal if and only if $[\![C[P]]\!] = [\![C[Q]]\!]$ for all contexts $C[\,]$ of base type, i.e. observational equality can be reformulated without any reference to an operational semantics.

The denotational semantics considered in the sequel will be *compositional* in the sense that from $[\![P]\!] = [\![Q]\!]$ it follows that $[\![C[P]]\!] = [\![C[Q]]\!]$ for all contexts $C[\,]$ (not only those of base type). Thus, for compositional computationally adequate denotational semantics from $[\![P]\!] = [\![Q]\!]$ it follows that P and Q are observationally equal. Actually, this already entails

---

[5] One also might say that "the operational semantics is computationally adequate w.r.t. the denotational semantics" because the denotational semantics may be considered as conceptually prior to the operational semantics. One could enter an endless "philosophical" discussion on what comes first, the operational or the denotational semantics. The authors have a slight preference for the view that denotational semantics should be conceptually prior to operational semantics (the *What* comes before the *How*) being, however, aware of the fact that in practice operational semantics often comes before the denotational semantics.

completeness of the denotational semantics as if $[\![P]\!] = n = [\![\underline{n}]\!]$ then P and $\underline{n}$ are observationally equal from which it follows that $P{\Downarrow}\underline{n} \Leftrightarrow \underline{n}{\Downarrow}\underline{n}$ and, therefore, $P{\Downarrow}\underline{n}$ as $\underline{n}{\Downarrow}\underline{n}$ does hold anyway. Thus, under the assumption of correctness for a compositional denotational semantics computational adequacy is equivalent to the requirement that denotational equality entails observational equality.

Full Abstraction

For those people who think that operational semantics is prior to denotational semantics the notion of observational equality is more basic than denotational equality because the former can be formulated without reference to denotational semantics. From this point of view computational adequacy is sort of a "correctness criterion" as it guarantees that semantic equality entails the "real" observational equality (besides the even more basic requirement that denotation is an invariant of evaluation).

However, one might also require that denotational semantics is complete w.r.t. operational semantics in the sense that observational equality entails denotational equality, in which case one says that the denotational semantics is *fully abstract* w.r.t. the operational semantics. At first sight this may seem a bit weird because in a sense denotational semantics is more abstract than operational semantics as due to its extensional character it abstracts from intensional aspects such as syntax. However, observational equivalence—though defined *a priori* in operational terms—is more abstract than denotational equality under the assumption of computational adequacy guaranteeing that denotational equality entails observational equality. Accordingly, a fully abstract semantics induces a notion of denotational equality which is "as abstract as reasonably possible" where "reasonable" here means that terms are not identified if they can be distinguished by observations.

Notice, moreover, that under the assumption of computational adequacy full abstraction can be formulated without reference to operational semantics as follows: closed expressions P and Q (of the same type) are denotationally equal already if $C[P]$ and $C[Q]$ are denotationally equal for all contexts $C[\,]$ of base type. A denotational semantics satisfying this condition is fully abstract w.r.t. an operational semantics iff it is computationally adequate w.r.t. this operational semantics.

Whereas computational adequacy holds for almost all models of PCF this is not the case for full abstraction as exemplified by the (otherwise sort of canonical) Scott model. Though the Scott model (and, actually, also all other models considered in the literature) is fully abstract for closed expressions of first order types $\mathbf{nat}{\to}\mathbf{nat}{\to}\ldots{\to}\mathbf{nat}{\to}\mathbf{nat}$ full abstraction fails already for the second order type $(\mathbf{nat}{\to}\mathbf{nat}{\to}\mathbf{nat}){\to}\mathbf{nat}$.

However, the Scott model is fully abstract for an extension of PCF by a *parallel*, though deterministic, language construct $\mathsf{por} : \mathbf{nat}{\to}\mathbf{nat}{\to}\mathbf{nat}$, called "parallel or", which gives 0 as result if its first or its second argument equals 0, 1 if both arguments equal 1 and delivers $\bot$ as result in all other cases. This example illustrates quite forcefully the *relativity* of the notion of full abstraction w.r.t. the language under consideration. The only reason why the Scott model fails to be fully abstract w.r.t. PCF is that it distinguishes closed expressions E_1 and E_2 of the type $(\mathbf{nat}{\to}\mathbf{nat}{\to}\mathbf{nat}){\to}\mathbf{nat}$ although these cannot be distinguished by program contexts $C[\,]$ expressible in the language of PCF. However, E_1 and E_2 can be distinguished by the context $[\,](\mathsf{por})$. In other words whether a denotational semantics is fully abstract for a language strongly depends on the expressiveness of this very language. Accordingly, a lack of full abstraction can be repaired in two possible, but different ways

(1) keep the model under consideration but extend the language in a way such that the extension can be interpreted in the given model and denotationally different terms can be separated by program contexts expressible in the extended language (e.g. keep the Scott model but extend PCF by por) *or*

(2) keep the language and alter the model to one which is fully abstract for the given language.

Whether one prefers (1) or (2) depends on whether one gives preference to the model or to the syntax, i.e. the language under consideration. A mathematician's typical attitude would be (1), i.e. to extend the language in a way that it can grasp more aspects of the model, simply because he is interested in the structure and the language is only a secondary means for communication. However, (even) a (theoretical) computer scientist's attitude is more reflected by (2) because for him the language under consideration is the primary concern whereas the model is just regarded as a tool for analyzing the language. Of course, one could now enter an endless discussion on which attitude is the more correct or more adequate one. The authors' opinion rather is that each single attitude when taken absolutely is somewhat disputable as (i) why shouldn't one take into account various different models instead of stubbornly insisting on a particular "pet model" and (ii) why should one take the language under consideration as absolute because even if one wants to exclude por for reasons of efficiency why shouldn't one allow[6] the observer to use it?

Instead of giving a preference to (1) or (2) we will present both approaches. We will show that extending PCF by por will render the Scott model fully abstract and we will present a refinement of the Scott model, the so-called *sequential domains*, giving rise to a fully abstract model for PCF which we consider as a final solution to a—or possibly *the*—most influential open problem in semantic research in the period 1975–2000. The solution via sequential domains is mainly known under the name "relational approach" because domains are endowed with (a lot of) additional relational structure which functions between sequential domains are required to preserve in addition to the usual continuity requirements of Scott's Domain Theory.

A competing and, actually, more influential approach is via *game semantics* where types are interpreted as games and programs as strategies. However, this kind of models is never extensional and, accordingly, not fully abstract for PCF as by Milner's Context lemma extensional equality entails observational equality. However, the "extensional collapse" of games models turns out as fully abstract for PCF. But this also holds for the term model of PCF and in this respect the game semantic approach cannot really be considered as a genuine solution of the full abstraction problem at least according to its traditional understanding. However, certain variations of game semantics are most appropriate for constructing fully abstract models for non-functional extensions of PCF, e.g. by control operators or references, as for such extensions the term models obtained by factorisation w.r.t. observational equivalence are *not extensional* anymore and, therefore, the inherently extensional approach via domains is not applicable anymore.

Notice that there is also a more liberal notion of sequentiality, namely the *strongly stable domains* of T. Ehrhard and A. Bucciarelli where, however, the ordering on function spaces is not pointwise anymore.

Universality

In the Scott model one can distinguish for every type σ a subset $C_\sigma \subseteq D_\sigma$ of *computable* elements without any reference to PCF-definability such that all PCF-definable elements of D_σ are already contained in C_σ. Now, if one has fixed such a semantic notion of computability for a model then there arises the question whether *all computable elements of the model do*

---

[6] as for example in cryptology where the attacker is usually assumed to employ as strong weapons as possible

arise as denotations of closed PCF *terms* in which case the model is called *universal*.[7]

A language universal for the Scott model can be obtained from PCF by adding por ("parallel or") and Plotkin's *continuous existential quantifier* $\exists$ of type $(\mathbf{nat} \rightarrow \mathbf{nat}) \rightarrow \mathbf{nat}$ which is defined as follows: $\exists(f) = 0$ if $f(n) = 0$ for some $n \in \mathbb{N}$, $\exists(f) = 1$ if $f(\bot) = 1$ and $\exists(f) = \bot$ in all other cases.

Notice, however, that $\exists$ cannot be implemented within PCF+por from which it follows that universality is a stronger requirement than full abstraction. But universality entails full abstraction as there is a theorem saying that a model of PCF is fully abstract iff all its "finite" elements are PCF definable and as these "finite" elements are subsumed by any reasonable notion of computability.

We conclude this introductory chapter by discussing the relevance of denotational semantics for **logics of programs**, i.e. calculi where properties of programs can be expressed and verfied.

First of all denotational models of programming languages are needed for defining validity of assertions about programs as can be expressed in a logic for this programming language. In case of PCF the family $(D_\sigma)_{\sigma \in \mathsf{Type}}$ provides the carriers for a many-sorted structure in which one can interpret the terms of the program logic LCF (Logic of Computable Functionals)[8] whose terms are expressions of the programming language PCF and whose formulas are constructed via the connectives and quantifiers of first order logic from atomic formulas $t_1 \sqsubseteq t_2$ stating that the meaning of t_1 is below the meaning of t_2 w.r.t. the information ordering as given by the denotational model. Notice, however, that the term language PCF is not first order as it contains a binding operator λ needed for explicit definitions of functions. However, this does not cause any problems for the interpretation of LCF. Instead of first order logic one might equally well consider higher

---

[7]Calling this property "universal" is in accordance with the common terminology where a programming language L is called "Turing universal" iff all partial recursive functions on $\mathbb{N}$ can be implemented by programs of L. The property "universal" as defined above is stronger since it requires that computable elements of *all* types can be implemented within the language under consideration. But in both cases "universal" means that one has already got an implementation for all possible computable elements (of a certain kind).

[8]The calculus LCF was introduced by D. Scott in an unpublished, but widely circulated and most influential manuscript dating back to 1967. In the 1970ies a proof assistant for LCF was implemented by R. Milner who for this very purpose developed and implemented the functional programming language ML (standing for "Meta-Language") whose refined versions SML and OCAML today constitute the most prominent typed call-by-value functional programming languages.

order logic over a model of PCF which has the advantage that higher order logic allows one to express inductively defined predicates which are most useful for the purposes of program verifiaction.

In principle one could interpret LCF also in the structure obtained by factorizing the closed PCF terms modulo observational equality. However, such a structure is not very easy to analyze as it is too concrete. Denotational models have the advantage that simple and strong proof principles like *fixpoint induction, computational induction* and *Park induction*, which are indispensible for reasoning about recursively defined functions and objects, can be easily verified for these models as they are actually derived from some obvious properties of these models.

Chapter 2

PCF and its Operational Semantics

In this chapter we introduce the prototypical functional programming language PCF together with its operational semantics.

The language PCF is a typed language whose set Type of types is defined inductively as follows

- the base type **nat** is a type and
- whenever σ and τ are types then $(\sigma\rightarrow\tau)$ is a type, too.

We often write ι for base type **nat** and $\sigma\rightarrow\tau$ instead of $(\sigma\rightarrow\tau)$ where $\rightarrow$ is understood as a right associative binary operation on Type meaning that e.g. $\sigma_1\rightarrow\sigma_2\rightarrow\sigma_3$ is understood as standing for $\sigma_1\rightarrow(\sigma_2\rightarrow\sigma_3)$. Due to the inductive definition of Type every type σ is of the form $\sigma_1\rightarrow\ldots\rightarrow\sigma_n\rightarrow\iota$ in a unique way.

As PCF terms may contain free variables we will define terms relative to *type contexts* where finitely many variables are declared together with their types, i.e. type contexts are expressions of the form

$$\Gamma \equiv x_1{:}\sigma_1, \ldots, x_n{:}\sigma_n$$

where the σ_i are types and the x_i are pairwise distinct variables. As variables cannot occur in type expressions the order of the single variable declarations $x_i{:}\sigma_i$ in Γ is irrelevant and, accordingly, we identify Γ with Γ' if the latter arises from the former by a permutation of the $x_i{:}\sigma_i$.

The valid judgements of the form

$$\Gamma \vdash M : \sigma \qquad (M \text{ is a term of type } \sigma \text{ in context } \Gamma)$$

are defined inductivly by the rules in Figure 2.1.

One easily shows by induction on the structure of derivations that whenever $\Gamma \vdash M : \sigma$ can be derived then $\pi(\Gamma) \vdash M : \sigma$ can be derived, too, for

Typing Rules for PCF

$$\frac{}{\Gamma, x{:}\sigma, \Delta \vdash x : \sigma} \qquad \frac{\Gamma, x{:}\sigma \vdash M : \tau}{\Gamma \vdash (\lambda x{:}\sigma.\, M) : \sigma {\to} \tau}$$

$$\frac{\Gamma \vdash M : \sigma {\to} \tau \quad \Gamma \vdash N : \sigma}{\Gamma \vdash M(N) : \tau} \qquad \frac{\Gamma \vdash M : \sigma {\to} \sigma}{\Gamma \vdash \mathsf{Y}_\sigma(M) : \sigma}$$

$$\frac{}{\Gamma \vdash \mathsf{zero} : \mathbf{nat}} \qquad \frac{\Gamma \vdash M : \mathbf{nat}}{\Gamma \vdash \mathsf{succ}(M) : \mathbf{nat}}$$

$$\frac{\Gamma \vdash M : \mathbf{nat}}{\Gamma \vdash \mathsf{pred}(M) : \mathbf{nat}} \qquad \frac{\Gamma \vdash M_i : \mathbf{nat} \quad (i{=}1,2,3)}{\Gamma \vdash \mathsf{ifz}(M_1, M_2, M_3) : \mathbf{nat}}$$

Figure 2.1   Typing rules for PCF

every permutation π of Γ.

As for every language construct of PCF there is precisely one typing rule one easily shows (Exercise!) that the σ with $\Gamma \vdash M : \sigma$ is determined uniquely by Γ and M. Thus, applying these typing rules backwards gives rise to a recursive *type checking* algorithm which given M and Γ computes the type σ with $\Gamma \vdash M : \sigma$ provided it exists and reports failure otherwise. (We invite the reader to test this algorithm for some simple examples!)

In the sequel we will not always stick to the "official" syntax of PCF terms as given by the typing rules. Often we write MN or (MN) instead of $M(N)$. In accordance with right-associativity of $\to$ we assume that application as given by juxtaposition is left-associative meaning that $M_1 \ldots M_n$ is read as $(\ldots (M_1 M_2) \ldots M_n)$ or $M_1(M_2) \ldots (M_n)$, respectively.

For variables bound by λ's we employ the usual convention of α-*conversion* according to which terms are considered as equal if they can be obtained from each other by an appropriate renaming of bound variables. Furthermore, when substituting term N for variable x in term M we first rename the bound variables of M in such a way that free variables of N will not get bound by *lambda*-abstractions in M, i.e. we employ so-called *capture-free substitution*.[1]

---

[1]These are the same conventions as usually employed for the quantifiers $\forall$ and $\exists$. The only difference is that quantifiers turn formulas into formulas whereas λ-abstraction

Before we define the operational semantics of PCF we introduce the notion of "raw terms" of PCF as given by the following grammar

$$M ::= \quad x \mid (\lambda x{:}\sigma.M) \mid M(M) \mid \mathsf{Y}_\sigma(M) \mid$$
$$\mathsf{zero} \mid \mathsf{succ}(M) \mid \mathsf{pred}(M) \mid \mathsf{ifz}(M, M, M)$$

in BNF form. Of course, not every raw term is typable as for example $\lambda x{:}\mathbf{nat}.x(x)$ where the first occurrence of x would have to be of functional type in order to render $x(x)$ well-typed.

We now present a "big step" semantics for PCF by inductively defining a binary relation $\Downarrow$ on raw terms via the rules exhibited in Figure 2.2 where $\underline{n}$ is the canonical *numeral* for the natural number n defined as $\underline{0} \equiv \mathsf{zero}$ and $\underline{k+1} \equiv \mathsf{succ}(\underline{k})$ by recursion on k.[2]

Bigstep Semantics for PCF

$$\frac{}{x \Downarrow x} \qquad\qquad\qquad \frac{}{\lambda x{:}\sigma.M \Downarrow \lambda x{:}\sigma.M}$$

$$\frac{M \Downarrow \lambda x{:}\sigma.E \quad E[N/x] \Downarrow V}{M(N) \Downarrow V} \qquad \frac{M(\mathsf{Y}_\sigma(M)) \Downarrow V}{\mathsf{Y}_\sigma(M) \Downarrow V}$$

$$\frac{}{\underline{0} \Downarrow \underline{0}} \qquad\qquad\qquad \frac{M \Downarrow \underline{n}}{\mathsf{succ}(M) \Downarrow \underline{n+1}}$$

$$\frac{M \Downarrow \underline{0}}{\mathsf{pred}(M) \Downarrow \underline{0}} \qquad\qquad \frac{M \Downarrow \underline{n+1}}{\mathsf{pred}(M) \Downarrow \underline{n}}$$

$$\frac{M \Downarrow \underline{0} \quad M_1 \Downarrow V}{\mathsf{ifz}(M, M_1, M_2) \Downarrow V} \qquad \frac{M \Downarrow \underline{n+1} \quad M_2 \Downarrow V}{\mathsf{ifz}(M, M_1, M_2) \Downarrow V}$$

Figure 2.2   Bigstep Semantics for PCF

Whenever $E \Downarrow V$ then V is a variable, a numeral or a λ-abstraction. It follows by induction on the structure of derivations of $E \Downarrow V$ that the free

turns terms into terms.

[2]Notice that in the literature one finds variants of PCF where instead of zero there are constants $\underline{n}$ for every natural number n. However, the same rules can be used for defining $\Downarrow$ inductively (albeit with a slightly different reading).

variables of V are contained in the free variables of E. Thus, if E is a closed expression and $E{\Downarrow}V$ then V is either a numeral or a λ-abstraction without free variables. Such terms are called (syntactic) *values* and one can see easily that for every such value V it holds that $V{\Downarrow}V$. Thus syntactic values are those terms V such that $M{\Downarrow}V$ can be derived for some closed term M. Notice that $\lambda x{:}\sigma.M$ is a value even if M is not a value, i.e. evaluation stops as soon as it has arrived at a functional abstraction. In our investigations of PCF we are mainly interested in closed terms and will hardly ever need the evaluation rule for variables. This is also the reason why we have not included variables into our definition of syntactic values.

Notice that with the exception of pred and ifz for each construct of PCF there is precisely one evaluation rule. In case of pred and ifz there are two rules which, however, do not overlap (in the sense that for every term at most one of these two rules is applicable). This observation gives rise to the following lemma.

Lemma 2.1   *The evaluation relation* $\Downarrow$ *is* deterministic, *i.e. whenever* $M{\Downarrow}V$ *and* $M{\Downarrow}W$ *then* $V \equiv W$.

Proof.   Straightforward induction on the structure of derivations of $M{\Downarrow}V$. (Exercise!) $\square$

Next we will show that evaluation preserves types, a property which is usually called *Subject Reduction*.

Theorem 2.2   (Subject Reduction)
If $\vdash M : \sigma$ *and* $M{\Downarrow}V$ *then* $\vdash V : \sigma$.

Proof.   Straightforward induction (Exercise!) on the structure of derivations of $M{\Downarrow}V$. $\square$

Thus, if M is a closed term of type **nat** and $M{\Downarrow}V$ then $V \equiv \underline{n}$ for some natural number n and if M is a closed term of type $\sigma{\to}\tau$ and $M{\Downarrow}V$ then $V \equiv \lambda x{:}\sigma.E$ for some E with $x{:}\sigma \vdash E : \tau$.

Often in the literature one can find definitions of PCF with a base type **bool** of boolean values included. In this case one adds the following term formation rules

$$\frac{\rule{2cm}{0.4pt}}{\Gamma \vdash \mathsf{true} : \mathbf{bool}} \qquad \frac{\rule{2cm}{0.4pt}}{\Gamma \vdash \mathsf{false} : \mathbf{bool}}$$

$$\frac{\Gamma \vdash M : \mathbf{bool} \quad \Gamma \vdash M_1 : \sigma \quad \Gamma \vdash M_2 : \sigma}{\Gamma \vdash \mathsf{cond}_\sigma(M, M_1, M_2) : \sigma} \ (\sigma \in \{\mathbf{nat}, \mathbf{bool}\})$$

together with the following evaluation rules

$$\overline{\mathsf{true} \Downarrow \mathsf{true}} \qquad \overline{\mathsf{false} \Downarrow \mathsf{false}}$$

$$\frac{M \Downarrow \mathsf{true} \quad M_1 \Downarrow V}{\mathsf{cond}_\sigma(M, M_1, M_2) \Downarrow V} \qquad \frac{M \Downarrow \mathsf{false} \quad M_2 \Downarrow V}{\mathsf{cond}_\sigma(M, M_1, M_2) \Downarrow V}$$

Notice that in this case ifz can be replaced by a predicate isz, i.e. $\mathsf{isz}(M)$ is a term of type **bool** whenever M is a term of type **nat** and $\mathsf{isz}(M)$ evaluates to true iff $M \Downarrow \underline{0}$ and to false iff $M \Downarrow \underline{n+1}$ for some natural number n. Using isz we can implement ifz by putting $\mathsf{ifz}(M, M_1, M_2) \equiv \mathsf{cond}_{\mathbf{nat}}(\mathsf{isz}(M), M_1, M_2)$.

However, this extension by boolean values is fairly redundant as we can simulate boolean values within **nat** coding, say, true by $\underline{0}$ and false by $\underline{1}$.

Next we present a "single step" semantics for PCF and show that it coincides with the big step semantics. The single step semantics is given by specifying a relation $\triangleright$ between terms (of the same type) where $M \triangleright N$ reads as "M reduces in one step to N". This reduction relation $\triangleright$ is defined inductively by the rules given in Figure 2.3.

Only the first six rules of Figure 2.3 specify proper computation steps. The purpose of the remaining four rules is to fix a *leftmost outermost* reduction strategy. These last four rules could be replaced by a single one, namely

$$\frac{M_1 \triangleright M_2}{E[M_1] \triangleright E[M_2]}$$

where E ranges over *evaluation contexts* defined by the grammar

$$E := [\,] \mid E(M) \mid \mathsf{succ}(E) \mid \mathsf{pred}(E) \mid \mathsf{ifz}(E, M_1, M_2)$$

in BNF form. As for every term M there is at most one evaluation context E such that $M \equiv E[N]$ and N is the left hand side of some valid reduction $N \triangleright N'$ it follows that the reduction relation $\triangleright$ is deterministic.

Let us write $\triangleright^*$ for the reflexive transitive closure of $\triangleright$. One can show (Exercise!) that $M \Downarrow V$ iff $M \triangleright^* V$ and V is a syntactic value[3]. For this purpose one verifies (Exercise!) that

(a) if $M \Downarrow V$ then $M \triangleright^* V$ and
(b) if $M \triangleright N$ then for all values V, if $N \Downarrow V$ then $M \Downarrow V$

---

[3]Notice that V is a value if there is no term N with $V \triangleright N$.

by induction on the structure of derivations of $M{\Downarrow}V$ and $M \vartriangleright N$, respectively. Applying (b) iteratively it follows that

(c) if $M \vartriangleright^* N$ then for all values V, if $N{\Downarrow}V$ then $M{\Downarrow}V$.

Then from (a) and (c) it follows immediately that $M{\Downarrow}V$ if and only if $M \vartriangleright^* V$ for all terms M and values V. Thus big step and small step semantics for PCF coincide. Of course, big step semantics is more abstract in the sense that it forgets about intermediary computation steps. That is the reason why we stick to big step semantics when studying the relation between operational and denotational semantics of PCF.

Smallstep Semantics for PCF

$$\overline{(\lambda x{:}\sigma.M)(N) \vartriangleright M[N/x]} \qquad \overline{\mathsf{Y}_\sigma(M) \vartriangleright M(\mathsf{Y}_\sigma(M))}$$

$$\overline{\mathsf{pred}(\underline{0}) \vartriangleright \underline{0}} \qquad \overline{\mathsf{pred}(\underline{n{+}1}) \vartriangleright \underline{n}}$$

$$\overline{\mathsf{ifz}(\underline{0}, M_1, M_2) \vartriangleright M_1} \qquad \overline{\mathsf{ifz}(\underline{n{+}1}, M_1, M_2) \vartriangleright M_2}$$

$$\frac{M_1 \vartriangleright M_2}{M_1(N) \vartriangleright M_2(N)} \qquad \frac{M_1 \vartriangleright M_2}{\mathsf{succ}(M_1) \vartriangleright \mathsf{succ}(M_2)}$$

$$\frac{M_1 \vartriangleright M_2}{\mathsf{pred}(M_1) \vartriangleright \mathsf{pred}(M_2)} \qquad \frac{M_1 \vartriangleright M_2}{\mathsf{ifz}(M_1, N_1, N_2) \vartriangleright \mathsf{ifz}(M_2, N_1, N_2)}$$

Figure 2.3   Small Step Semantics for PCF

The syntactic preorders $\precsim$ ***and*** $\lesssim$

For every type σ we write Prg_σ for the set $\{M \mid {\vdash}M{:}\sigma\}$ of closed PCF terms of type σ also called *programs of type* σ. Programs of base type will be simply called programs. By induction on the structure of σ we will now define preorders $\precsim_\sigma$ and $\lesssim_\sigma$ on Prg_σ.

For base type **nat** we define

$$M \sqsubseteq_{\mathbf{nat}} N \quad \text{iff} \quad \forall n{\in}\mathbb{N}.\ M{\Downarrow}\underline{n} \Rightarrow N{\Downarrow}\underline{n}$$

and for functional types $\sigma{\to}\tau$ we define

$$M \sqsubseteq_{\sigma\to\tau} N \quad \text{iff} \quad \forall P{\in}\mathsf{Prg}_\sigma.\ M(P) \sqsubseteq_\tau N(P)\ .$$

The relation $\sqsubseteq$ will be called "applicative approximation" and we leave it as an exercise(!) to the reader to verify that $\sqsubseteq_\sigma$ is actually a preorder on Prg_σ, i.e. that $\sqsubseteq_\sigma$ is reflexive and transitive. One easily shows that for types $\sigma \equiv \sigma_1{\to}\ldots{\to}\sigma_n{\to}\mathbf{nat}$ it holds that $M \sqsubseteq_\sigma N$ iff $M\vec{P} \sqsubseteq_{\mathbf{nat}} N\vec{P}$ for all $\vec{P} \in \mathsf{Prg}_{\sigma_1} \times \ldots \times \mathsf{Prg}_{\sigma_n}$ (where we write $M\vec{P}$ for $M(P_1)\ldots(P_n)$ if $\vec{P}$ is the n–tuple $\langle P_1, \ldots, P_n \rangle$).

The "observational approximation" ordering $\precsim_\sigma$ at type σ is defined as

$$M \precsim_\sigma N \quad \text{iff} \quad \forall P{\in}\mathsf{Prg}_{\sigma\to\mathbf{nat}}.\ P(M) \sqsubseteq_{\mathbf{nat}} P(N)$$

where the underlying intuition is that every "observation" which can be made about M can also be made about N. Obviously, from $M \precsim_\sigma N$ it follows that $M \sqsubseteq_\sigma N$ as in the latter one quantifies only over a restricted class of observations, namely those of the form $\lambda x{:}\sigma.\ x\vec{P}$.

The classical *Milner's Context Lemma* says that both orderings are actually the same. However, its proof requires some sophistication and mathematical machinery. Accordingly, we postpone it to a subsequent chapter.

It is straightforward to see that for computationally adequate models it holds that $M \precsim N$ whenever $[\![M]\!] \sqsubseteq [\![N]\!]$. The reverse implications holds only for fully abstract models which, however, are difficult to construct.

An Abstract Environment Machine for PCF

We now will describe an abstract machine for evaluating PCF terms in order to give an idea of how functional languages can be implemented on traditional von Neumann machines.

At first sight one might be inclined to directly implement the small step semantics considered above, i.e. to implement the partial function on terms whose graph is the reduction relation $\triangleright$. However, this is not very efficient since replacing $(\lambda x.M)(N)$ by $M[N/x]$ is somewhat costy if there are many free occurrences of x in M which is in conflict with the intuitive requirement that single steps in a computation process should all be simple and change state only in a very local manner.

The key idea of an *environment machine* is to postpone the possibly costy operation of substitution as long as possible. For this reason the machine manipulates so-called *closures* which are pairs $[M, e]$ where M is a term and e is an *environment*, i.e. a finite function from variables to closures.

The syntax of untyped PCF terms is given by the grammar

$$M ::= x \mid \lambda x.M \mid M(M) \mid \mathsf{Y}(M) \mid \mathsf{zero} \mid \mathsf{succ}(M) \mid \mathsf{pred}(M) \mid \mathsf{ifz}(M, M, M)$$

in BNF form. We consider untyped PCF terms as type information is irrelevant for the computation process.

We write $\emptyset$ for the empty environment and $e[x{:=}c]$ for the environment which behaves like e for variables different from x and sends x to the closure c. We also write $\mathsf{dom}(e)$ for the finite set of variables to which e assigns a closure. Obviously, we have $\mathsf{dom}(e[x{:=}c]) = \mathsf{dom}(e) \cup \{x\}$.

The *states* of the abstract machines will be pairs $\langle c, S \rangle$ where c is a closure and S is a *stack* or *continuation* which are defined by the following grammar

$$S ::= \mathsf{stop} \mid \mathsf{arg}(c, S) \mid \mathsf{succ}(S) \mid \mathsf{pred}(S) \mid \mathsf{ifz}(M, M, e, S)$$

in BNF form.

Finally the transition rules of the Abstract Environment Machine for PCF are given in Figure 2.4.

We now try to reveal the intuition behind the various transition rules.

The first three rules are sufficient to compute weak head normal forms of terms of untyped λ-calculus. Recall that a weak head normal form is either a variable or a λ-abstraction. For this fragment the continuations are stacks where arg takes a closure c and pushes it on stack S. When an application term has to be evaluated its argument together with the current environment is pushed on the stack. This is iterated until one lands in case (1) or (2). In the first case the variable x is replaced by the closure $e(x)$ where e is the current environment provided $e(x)$ is defined and otherwise we have found the head variable of the term. A λ-expression $\lambda x.M$ under current environment e is evaluated by evaluating its body M in the environment $e[x{:=}c]$ where c is the closure on top of the current stack. If the current stack is empty then $(\lambda x.M)[e]$ is the weak head normal form.

Rule (4) extends this to general recursion as given by Y. Thus, in order to evaluate $\mathsf{Y}(M)$ under the environment e evaluate $M(\mathsf{Y}(M))$ under environment e which, however, by (3) is evaluated as follows: push the argument $\mathsf{Y}(M)$ together with e on the stack and then evaluate M w.r.t. e. Rule (4) has the same effect but achieves it in one single step.

Transition Rules of the Abstract Environment Machine

(1) $\langle [x, e], S \rangle \to \langle e(x), S \rangle$ if $x \in \mathsf{dom}(e)$

(2) $\langle [\lambda x.M, e], \mathsf{arg}(c, S) \rangle \to \langle [M, e[x{:=}c]], S \rangle$

(3) $\langle [M(N), e], S \rangle \to \langle [M, e], \mathsf{arg}([N, e], S) \rangle$

(4) $\langle [\mathsf{Y}(M), e], S \rangle \to \langle [M, e], \mathsf{arg}([\mathsf{Y}(M), e], S) \rangle$

(5) $\langle [\mathsf{succ}(M), e], S \rangle \to \langle [M, e], \mathsf{succ}(S) \rangle$

$\quad\;\;\langle [\underline{n}, e], \mathsf{succ}(S) \rangle \to \langle [\underline{n{+}1}, e], S \rangle$

(6) $\langle [\mathsf{pred}(M), e], S \rangle \to \langle [M, e], \mathsf{pred}(S) \rangle$

$\quad\;\;\langle [\underline{0}, e], \mathsf{pred}(S) \rangle \to \langle [\underline{0}, e], S \rangle$

$\quad\;\;\langle [\underline{n{+}1}, e], \mathsf{pred}(S) \rangle \to \langle [\underline{n}, e], S \rangle$

(7) $\langle [\mathsf{ifz}(M, N_1, N_2), e], S \rangle \to \langle [M, e], \mathsf{ifz}(N_1, N_2, e, S) \rangle$

$\quad\;\;\langle [\underline{0}, e'], \mathsf{ifz}(N_1, N_2, e, S) \rangle \to \langle [N_1, e], S \rangle$

$\quad\;\;\langle [\underline{n{+}1}, e'], \mathsf{ifz}(N_1, N_2, e, S) \rangle \to \langle [N_2, e], S \rangle$

Figure 2.4   Abstract Environment Machine for PCF

Whereas application follows a call-by-name strategy expressions of the form $\mathsf{succ}(M)$ or $\mathsf{pred}(M)$ are evaluated following a call-by-value strategy. Therefore it is not appropriate to push the argument M together with the current environment e on the current stack S. Instead one evaluates M w.r.t. e and the stack $\mathsf{succ}(S)$. When this evaluation has resulted in the closure $[\underline{n}, e']$ (tacitly assuming that the current stack is again $\mathsf{succ}(S)$) then evaluate $[\underline{n{+}1}, e']$ w.r.t. the original stack S. For pred the procedure is analogous.

As ifz is call-by-value in its first argument when evaluating an expression of the form $\mathsf{ifz}(M, N_1, N_2)$ w.r.t. environment e and stack S one first has to evaluate M w.r.t. e but relative to the stack $\mathsf{ifz}(N_1, N_2, e, S)$ which keeps the information how to continue when $[M, e]$ has been evaluated to a numeral.[4] Depending on whether this numeral is 0 or greater 0 one proceeds by evaluating N_1 w.r.t. e and S or by evaluating N_1 w.r.t. e and S.

The formal verification of the correctness of our environment machine is somewhat delicate and we omit it as it isn't the main concern of this course but rather of a course on implementations of functional programming languages.

---

[4]That is the reason why stacks are often called "continuations". They tell us how to "continue" after an intermediary result has been found.

Chapter 3

The Scott Model of PCF

In this chapter we introduce the kind of structures within which Dana Scott has interpreted the language PCF (and its logic LCF). (See [Scott 1969] for a reprint of a widely circulated "underground" paper from 1969 where this interpretation was presented the first time.) But before we will discuss the general form of a denotational semantics for PCF and try to motivate some of the structural requirements we impose.

A denotational semantics for PCF associates with every type σ a so-called *domain* D_σ and with every term $x_1{:}\sigma_1, \ldots, x_n{:}\sigma_n \vdash M : \sigma$ a function

$$\llbracket x_1{:}\sigma_1, \ldots, x_n{:}\sigma_n \vdash M : \sigma \rrbracket : D_{\sigma_1} \times \cdots \times D_{\sigma_n} \to D_\sigma$$

assuming that cartesian products of domains exist. In case M is a closed term (i.e. $n{=}0$) we have $\llbracket \vdash M : \sigma \rrbracket : 1 \to D_\sigma$ where 1 stands for the empty product containing just the empty tuple $\langle \rangle$ as its single element.

We have tacitly assumed that domains are sets (and that their finite products are defined as for sets). But notice that one must *not* interpret $D_{\sigma \to \tau}$ as the set of *all* functions from D_σ to D_τ as then one would run into problems with interpreting the fixpoint operators Y_σ as their interpretation would have to associate with every $f \in D_{\sigma \to \sigma}$, i.e. with every function f from D_σ to D_σ, a fixpoint of f, i.e. a $Y_\sigma(f) \in D_\sigma$ satisfying the fixpoint equation $Y_\sigma(f) = f(Y_\sigma(f))$, and such a fixpoint need not exist in general (e.g. if f is a fixpoint free permutation of the set D_σ). The solution to this problem is to endow the domains D_σ with additional structure and to require that $D_{\sigma \to \tau}$ consists of all maps from D_σ to D_τ which do preserve this structure. Of course, we then have to endow this set also with an appropriate structure of that kind.

The question now is to identify what is an appropriate structure to impose on domains which serve the purpose of interpreting PCF (or other programming languages). In particular, this kind of structure should not be arbitrary but rather well motivated by operational phenomena. Well, in the previous chapter we have seen that for every type σ one can define the preorder $\lesssim_\sigma$ on the set Prg_σ of programs of type σ where $M \lesssim_\sigma N$ means that N contains all the information of M and possibly more. By analogy this suggests to endow the domains with a partial ordering called "information ordering".

If one factors the closed terms of type **nat** by $\lesssim_{\mathbf{nat}}$ one obtains the poset (i.e. partially ordered set) N whose underlying set is $\mathbb{N} \cup \{\bot\}$ where $\bot$ (read "bottom") is a distinguished object (not contained in $\mathbb{N}$) representing *nontermination* or *divergence*. Actually, for every type σ there is a closed term $\Omega_\sigma \equiv \mathsf{Y}_\sigma(\lambda x{:}\sigma.x)$ with $\Omega_\sigma \lesssim_\sigma M$ for all $M \in \mathsf{Prg}_\sigma$. Thus, we require every domain D_σ to be endowed with a partial order $\sqsubseteq_\sigma$ and to contain a least element $\bot_{D_\sigma}$. As $\lesssim$ coincides with $\lesssim$ by Milner's Context Lemma[1] every program P of type $\sigma{\to}\tau$ preserves $\lesssim$ as it obviously preserves $\lesssim$. This leads us to the requirement that the functions $f \in D_{\sigma\to\tau}$ should be monotonic, i.e. preserve the partial order $\sqsubseteq$. As by definition $M \lesssim_{\sigma\to\tau} N$ iff $M(P) \lesssim N(P)$ for all programs P of type σ it appears as natural to define the partial order $\sqsubseteq$ on $D_{\sigma\to\tau}$ as the *pointwise ordering* according to which $f \sqsubseteq g$ iff $\forall d{\in}D_\sigma.\, f(d) \sqsubseteq g(d)$.

However, it is not sufficient to require that domains are partial orders with a least element and functions between them have to be monotonic because this does not yet guarantee the existence of fixpoints. Consider for example the set $\mathbb{N}$ of natural numbers under their usual ordering $\leq$ for which the successor function $f : \mathbb{N} \to \mathbb{N} : n \mapsto n{+}1$ is surely monotonic but obviously has no fixpoint.

This problem can be overcome by postulating that every domain has suprema of chains and functions between domains are not only monotonic but have to preserve also suprema of chains. Such functions between domains are called "(Scott) continuous". This has the advantage that for every domain D every continuous function $f : D \to D$ has a least fixpoint $\mu(f)$ which is obtained as the supremum of the chain

$$\bot \sqsubseteq f(\bot) \sqsubseteq f^2(\bot) \sqsubseteq \cdots \sqsubseteq f^n(\bot) \sqsubseteq \cdots$$

That $\mu(f)$ is actually a fixpoint of f follows from continuity of f as we have $f(\bigsqcup_n f^n(\bot)) = \bigsqcup_n f(f^n(\bot)) = \bigsqcup_n f^n(\bot)$. That $\mu(f)$ is actually the *least* fixpoint of f can be seen as follows: if $d = f(d)$ then by induction one

---

[1] which still has to be proved but may well serve the purpose of motivation!

easily shows that $f^n(\bot) \sqsubseteq d$ for all $n \in \mathbb{N}$ and thus $\mu(f) \sqsubseteq d$ since $\mu(f)$ is the supremum of the $f^n(\bot)$ which are bounded by d.

Summarizing we notice that the above considerations suggest that

- domains are partially ordered sets with a least element and suprema for all (weakly) increasing chains and
- functions between domains should preserve the partial ordering and suprema of (weakly) increasing chains.

One might be inclined to require functions between domains to preserve also least elements. This, however, would have the most undesirable consequences that (1) every constant map has value $\bot$ and (2) the least fixpoint of every endomap is $\bot$ rendering all recursive definitions trivial.

In the following for aesthetical reasons we require not only existence and preservation of suprema of chains but existence and preservation of suprema of so-called *directed* sets.

In the next two sections we develop some basic domain theory and then introduce the Scott model of PCF.

3.1   Basic Domain Theory

Definition 3.1   A *partial order (poset)* on a set D is a binary relation $\sqsubseteq_D \subseteq D \times D$ satisfying the following conditions

(reflexive)	$x \sqsubseteq_D x$
(transitive)	$x \sqsubseteq_D z$ whenever $x \sqsubseteq_D y$ and $y \sqsubseteq_D z$
(antisymmetric)	$x = y$ whenever $x \sqsubseteq_D y$ and $y \sqsubseteq_D x$.

A reflexive and transitive relation is called a *preorder*.

If $(D_1, \sqsubseteq_{D_1})$ and $(D_2, \sqsubseteq_{D_2})$ are preorders then a function $f : D_1 \to D_2$ is called *monotonic* iff $f(x) \sqsubseteq_{D_2} f(y)$ whenever $x \sqsubseteq_{D_1} y$. $\Diamond$

Obviously, monotonic maps are closed under composition and the identity function $\mathrm{id}_D : D \to D : d \mapsto d$ is a monotonic map from $(D, \sqsubseteq_D)$ to itself.

Definition 3.2   Let $(A, \sqsubseteq)$ be a poset. A subset $X \subseteq A$ is called *directed* iff every finite subset X_0 of X has an upper bound in X, i.e.

$$\forall X_0 \subseteq_{\mathsf{fin}} X. \exists y \in X. \forall x \in X_0.\ x \sqsubseteq y\ .$$

Thus, a directed set X is always nonempty because the empty set $\emptyset \subseteq_{\text{fin}} X$ has an upper bound in X.

A partial order $(A, \sqsubseteq)$ is called *predomain* or *complete partial order (cpo)* iff every directed subset of A has a least upper bound. A predomain $(A, \sqsubseteq)$ is called a *domain* or *pcpo (pointed cpo)* iff it has a least element $\bot$.

Let $(A_1, \sqsubseteq_{A_1})$ and $(A_2, \sqsubseteq_{A_2})$ be cpo's. A function from $(A_1, \sqsubseteq_{A_1})$ to $(A_2, \sqsubseteq_{A_2})$ is called *(Scott) continuous* iff it preserves suprema of directed sets, i.e.

$$f(\bigsqcup X) = \bigsqcup f(X)$$

for all directed $X \subseteq A_1$. A function between domains is called *strict* iff it preserves least elements. $\Diamond$

It is a straightforward exercise(!) to show that continuous functions between predomains are always monotonic.

Theorem 3.3   *Let $(A_i \mid i \in I)$ be a family of predomains. Then their product $\prod_{i \in I} A_i$ is a predomain under the componentwise ordering and the projections $\pi_i : \prod_{i \in I} A_i \to A_i$ are Scott continuous. If, moreover, all A_i are domains then so is their product $\prod_{i \in I} A_i$.*

If $(f : B \to A_i \mid i \in I)$ is a family of continuous maps between predomains then there is a unique continuous function $f : B \to \prod_{i \in I} A_i$ with

$$\pi_i \circ f = f_i$$

for all $i \in I$.

Proof.   Straightforward exercise! $\square$

Lemma 3.4   *Let A_1, A_2 and A_3 be cpos. Then a function $f : A_1 \times A_2 \to A_3$ is continuous iff it is continuous in each argument.*

Proof.   The implication from left to right is obvious.

For the reverse direction suppose that f is continuous in each argument. For showing that f is continuous consider an arbitrary directed subset $X \subseteq A_1 \times A_2$. Then for $i=1, 2$ the sets $X_i := \pi_i(X)$ are directed in A_i. Obviously, we have $\bigsqcup X = (\bigsqcup X_1, \bigsqcup X_2)$. As f is monotonic it suffices to show that

$$f(\bigsqcup X) \sqsubseteq \bigsqcup_{x \in X} f(x)$$

Suppose $z \sqsupseteq \bigsqcup_{x \in X} f(x)$. Then $z \sqsupseteq f(x_1, x_2)$ for all $x_1 \in X_1$ and $x_2 \in X_2$ (as if $(x_1, x'_2) \in X$ and $(x'_1, x_2) \in X$ then by directedness of X there is a $(y_1, y_2) \in X$ with $(y_1, y_2) \sqsupseteq (x_1, x_2'), (x_1', x_2))$. Thus, for all $x_1 \in X_1$ we have $z \sqsupseteq f(x_1, \bigsqcup X_2)$ as f is continuous in its second argument. Accordingly, as f is continuous also in its first argument we conclude that $z \sqsupseteq f(\bigsqcup X_1, \bigsqcup X_2) = f(\bigsqcup X)$ as desired.   □

Next we show that there are appropriate *function spaces* or *exponentials* in the category of predomains and continuous maps.

Theorem 3.5   *Let A_1 and A_2 be cpo's. Then the set $A_2^{A_1} = [A_1 \to A_2]$ of all Scott continuous maps from A_1 to A_2 is itself a cpo when ordered pointwise, i.e. when defining*

$$f \sqsubseteq g \qquad iff \qquad \forall a \in A_1.\ f(a) \sqsubseteq g(a)$$

for Scott continuous functions f and g.

Proof.   Let F be a directed subset of $[A_1 \to A_2]$. We show that its supremum $\bigsqcup F$ is given by the function g with

$$g(a) = \bigsqcup_{f \in F} f(a)$$

for $a \in A_1$. Notice that $g(a)$ is always defined because $\{f(a) \mid f \in F\}$ is directed. Obviously, the map g is the supremum of F provided g is continuous. It is easy to see that g is monotonic. Thus, for showing the continuity of g assume that X is a directed subset of A_1. As g is monotonic it suffices to show that

$$g(\bigsqcup X) \sqsubseteq \bigsqcup g(X)$$

For this purpose assume that $z \sqsupseteq \bigsqcup g(X)$, i.e. $z \sqsupseteq g(x)$ for all $x \in X$. Then z is also an upper bound for $\{f(x) \mid f \in F, x \in X\}$. Thus, for all $f \in F$ we have

$$z \sqsupseteq \bigsqcup f(X) = f(\bigsqcup X)$$

as f is continuous. Accordingly, the element z is also an upper bound of $g(\bigsqcup X)$ as desired.   □

As the *evaluation map*

$$\mathsf{ev} : [A_1 \to A_2] \times A_1 \to A_2 : (f, a) \mapsto f(a)$$

is continuous in each argument (exercise!) it follows by Lemma 3.4 that ev itself is continuous.

Theorem 3.6   *Let A, B and C be predomains. Then for every Scott continuous function $f : C \times A \to B$ there exists a unique Scott continuous function $g : C \to [A \to B]$ with*

$$g(z)(x) = f(z, x)$$

for all $x \in A$ and $z \in C$.

Proof.   Obviously, the function g is uniquely determined by the require-ment that $g(z)(x) = f(z, x)$ for all $x \in A$ and $z \in C$. As $g(z) = f(z, -)$ is continuous for all $z \in C$ it remains to show that g is continuous. For this purpose assume that Z is a directed subset of C. But then we have for all $x \in X$

$$g\Big(\bigsqcup Z\Big)(x) = f\Big(\bigsqcup Z, x\Big) = \bigsqcup_{z \in Z} f(z, x) = \Big(\bigsqcup_{z \in Z} g(z)\Big)(x)$$

where the last equality follows from the fact that directed suprema in $[A \to B]$ are constructed pointwise (see proof of Theorem 3.5). Thus, we have $g(\bigsqcup Z) = \bigsqcup_{z \in Z} g(z)$ as desired.   □

The claim of the previous theorem may be formulated more abstractly as follows: for every continuous $f : C \times A \to B$ there is a unique continuous $g : C \to [A \to B]$ such that the following diagram commutes

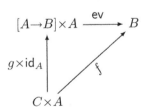

where $(g \times \mathrm{id}_A)(c, a) = (g(c), a)$. This requirement makes sense in every category with (binary) cartesian products and characterises the *exponential* $[A \to B]$ uniquely up to isomorphism. A category with finite products where for all objects A and B the exponential $[A \to B]$ exists is usually called *cartesian closed* (see e.g. [Scott 1980]).

One often writes $\Lambda(f)$ for the unique map g with $f = \mathrm{ev} \circ (g \times \mathrm{id}_A)$. We will see later that projections, ev and Λ provide enough structure for interpreting the simply typed λ-calculus in the category of predomains and continuous maps (and, actually, in an arbitrary cartesian closed category).

But now we dicuss fixpoints and fixpoint operators for domains.

Theorem 3.7   *Let D be a domain and $f : D \to D$ be continuous. Then the supremum*

$$\mu(f) = \bigsqcup_{n \in \mathbb{N}} f^n(\bot)$$

exists and satisfies the conditions

(1) $\mu(f) = f(\mu(f))$ *and*
(2) $\mu(f) \sqsubseteq d$ *whenever $f(d) \sqsubseteq d$.*

Thus, in particular $\mu(f)$ is the least *fixpoint of f.*

Proof.   First we show by induction on n that $f^n(\bot) \sqsubseteq f^{n+1}(\bot)$. Obviously, we have $f^0(\bot) = \bot \sqsubseteq f(\bot) = f^1(\bot)$ as $\bot$ is the least element of D. If $f^n(\bot) \sqsubseteq f^{n+1}(\bot)$ then $f^{n+1}(\bot) = f(f^n(\bot)) \sqsubseteq f(f^{n+1}(\bot)) = f^{n+2}(\bot)$ as f is monotonic. Thus $\mu(f) = \bigsqcup_{n \in \mathbb{N}} f^n(\bot)$ exists because directed sets have suprema in D. The element $\mu(f)$ is a fixpoint of f as we have

$$f(\mu(f)) = f\Big(\bigsqcup_{n \in \mathbb{N}} f^n(\bot) \Big) = \bigsqcup_{n \in \mathbb{N}} f(f^n(\bot)) = \bigsqcup_{n \in \mathbb{N}} f^{n+1}(\bot) = \mu(f)$$

where the second equality intrinsically makes use of continuity of f.

For the second claim suppose that $f(d) \sqsubseteq d$. We show by induction that $f^n(\bot) \sqsubseteq d$. Of course, we have $f^0(\bot) = \bot \sqsubseteq d$ as $\bot$ is the least element. If $f^n(\bot) \sqsubseteq d$ then $f^{n+1}(\bot) = f(f^n(\bot)) \sqsubseteq f(d) \sqsubseteq d$. Thus, it follows that $\mu(f) = \bigsqcup_{n \in \mathbb{N}} f^n(\bot) \sqsubseteq d$. That $\mu(f)$ is the least fixpoint follows immediately from the fact that $\mu(f)$ is below all *prefixpoints* $f(d) \sqsubseteq d$. □

Obviously, for arbitrary predomains A not every continuous endofunction $f : A \to A$ will have a fixpoint as this is wrong for sets and those live within predomains as the discrete partial orders.

By the previous theorem there is a function μ from $[D \to D]$ to D sending continuous f to their least fixpoint. One could show directly that μ is continuous, i.e. preserves suprema of directed sets. However, the following proof is much nicer.

Theorem 3.8   *Let D be a domain and $\Phi : [[D \to D] \to D] \to [[D \to D] \to D]$ the continuous operator with*

$$\Phi(F)(f) = f(F(f))$$

for $F \in [[D \to D] \to D]$ and $f \in [D \to D]$. The fixpoints of Φ are the continuous fixpoint operators on D and μ is the least fixpoint of Φ. Thus, the least fixpoint operator μ is continuous.

Proof.   First observe that

$$\Psi : [[D{\to}D]{\to}D]{\times}[D{\to}D] \to D : (F, f) \mapsto f(F(f))$$

is continuous in each argument (exercise!)   and thus continuous by Lemma 3.4. The operator Φ is continuous as one easily sees that $\Phi = \Lambda(\Psi)$.

A continuous function $F \in [[D{\to}D]{\to}D]$ is a fixpoint operator iff $F(f)$ is a fixpoint of f for all $f \in [D{\to}D]$, i.e. $F(f) = f(F(f))$ for all $f \in [D{\to}D]$, i.e. iff $F = \Phi(F)$. Thus, the fixpoints of Φ are precisely the continuous fixpoint operators.

The least fixpoint of Φ is μ because for all $n \in \mathbb{N}$ we have

$$\Phi^n(\bot)(f) = f^n(\bot)$$

for all $f \in [D{\to}D]$ as the following inductive argument shows. The claim is obvious for $n=0$ as $\Phi^0(\bot)(f) = \bot(f) = \bot = f^0(\bot)$. Suppose as induction hypothesis that $\Phi^n(\bot)(f) = f^n(\bot)$ for all $f \in [D{\to}D]$. Then for all $f \in [D{\to}D]$ we have

$$\Phi^{n+1}(\bot)(f) = \Phi(\Phi^n(\bot))(f) = f(\Phi^n(\bot)(f)) =_{(\mathrm{IH})} f(f^n(\bot)) = f^{n+1}(\bot)$$

proving the induction step. $\square$

Later we will interpret Y_σ of PCF as the least fixpoint operator for D_σ. The previous theorem guarantees that recursive definitions in PCF via the recursion operators Y_σ will not lead out of the world of Scott continuous functions which is indispensible for further, i.e. iterated, applications of Y.

"Induction Principles" for Least Fixpoints

As the least fixpoint operators of PCF are essential for writing nontrivial programs it is most desirable to have reasoning principles available for proving properties of least fixpoints. As some of these are *formally* analogous to proper induction principles it has become customary to call them "induction principles" though they do not verify that a certain property holds for all elements of some domain but rather that a property holds for a particular element, namely the least fixpoint of some given function.

Alas, most reasoning principles do not apply to arbitrary properties of domains. In the following definition we introduce a class of predicates on domains for which the subsequent induction principles are "admissible".

Definition 3.9   A subset P of a predomain A is called an *admissible predicate on A* iff P is closed under suprema of directed sets, i.e. $P(\bigsqcup X)$

for all directed $X \subseteq P$.   ◇

Notice that we often write $P(d)$ for $d \in P$ as we identify predicates with subsets. The first and most general reasoning principle is "computational induction".

Theorem 3.10 (Computational Induction)
Let D be a domain, $f : D \to D$ a continuous function and $P \subseteq D$ an admissible predicate on D. Then $P(\mu(f))$ whenever $P(f^n(\bot))$ for all $n \in \mathbb{N}$.

Proof.   As by assumption all elements of the directed set $\{f^n(\bot) \mid n \in \mathbb{N}\}$ are in P its supremum $\mu(f)$ is in P, too, as P is admissible.   □

The following immediate consequence called "fixpoint induction" is often easier to use.

Theorem 3.11   (Fixpoint Induction)
Let D be a domain, $f : D \to D$ a continuous function and $P \subseteq D$ an admissible predicate on D.
    Then $P(\mu(f))$ whenever $P(\bot)$ and $\forall x \in D. P(x) \Rightarrow P(f(x))$.

Proof.   From the premises $P(\bot)$ and $\forall x \in D. P(x) \Rightarrow P(f(x))$ one easily shows by ordinary induction on $\mathbb{N}$ that $\forall n \in \mathbb{N}. P(f^n(\bot))$ from which it follows by Theorem 3.10 that $P(\mu(f))$.   □

Notice the formal analogy of the structure of premises in ordinary induction

$$P(0) \wedge (\forall x. P(x) \Rightarrow P(\mathsf{succ}(x))) \Rightarrow \forall x. P(x)$$

and fixpoint induction

$$P(\bot) \wedge (\forall x. P(x) \Rightarrow P(f(x))) \Rightarrow P(\mu(f))$$

which was the reason for calling Theorem 3.11 "fixpoint induction".
    Finally we mention a proof principle due to David Park which is useful for showing that recursively defined functions diverge for some arguments.

Theorem 3.12   (Park Induction)
Let D be a domain and $f : D \to D$ a continuous function. Then $\mu(f) \sqsubseteq d$ whenever $f(d) \sqsubseteq d$.

Proof.   This is just Theorem 3.7 (2).   □

Now we have accumulated sufficiently much basic Domain Theory for introducing Scott's famous *domain model for* PCF.

3.2   Domain Model of PCF

We first describe the domains interpreting PCF types. For this purpose we need the following definition.

Definition 3.13   Let X be a set. Then $X_\perp$ is the poset whose underlying set of elements is $X \cup \{\perp\}$ where $\perp \notin X$ and which is partially ordered by

$$x \sqsubseteq y \quad \text{iff} \quad x = \perp \vee x = y \ .$$

The element $\perp$ is a fresh least element and the elements of X are all incomparable w.r.t. this ordering.   ◇

Now the domains D_σ associated with PCF types σ are defined inductively as follows

$$D_{\mathbf{nat}} = N \text{ where } N = \mathbb{N}_\perp \text{ and}$$
$$D_{\sigma \to \tau} = [D_\sigma \to D_\tau].$$

Notice that all D_σ contain a least element $\perp$. If one considers the extension of PCF by Boolean values then one puts $D_{\mathbf{bool}} = \mathbb{B}_\perp$ where $\mathbb{B} = \{\text{true}, \text{false}\}$ is the set of truth values.

If $\Gamma \equiv x_1{:}\sigma_1, \ldots, x_n{:}\sigma_n$ then we define $\llbracket \Gamma \rrbracket$ as $D_{\sigma_1} \times \ldots \times D_{\sigma_n}$. A term in context $\Gamma \vdash M : \tau$ will be interpreted as a function

$$\llbracket \Gamma \vdash M \rrbracket : \llbracket \Gamma \rrbracket \to D_\tau$$

which is required to be Scott continuous. The definition of the interpretation of terms in context proceeds by recursion over the structure of derivations as in the following definition.

Definition 3.14   The interpretation of terms in contexts is given by the following recursive clauses

$$\llbracket x_1{:}\sigma_1, \ldots, x_n{:}\sigma_n \vdash x_i \rrbracket (d_1, \ldots, d_n) = d_i$$
$$\llbracket \Gamma \vdash \lambda x{:}\sigma. M \rrbracket = \Lambda(\llbracket \Gamma, x{:}\sigma \vdash M \rrbracket)$$
$$\text{i.e. } \llbracket \Gamma \vdash \lambda x{:}\sigma. M \rrbracket(\vec{d})(d) = \llbracket \Gamma, x{:}\sigma \vdash M \rrbracket(\vec{d}, d)$$
$$\llbracket \Gamma \vdash M(N) \rrbracket(\vec{d}) = \mathsf{ev}(\llbracket \Gamma \vdash M \rrbracket(\vec{d}), \llbracket \Gamma \vdash N \rrbracket(\vec{d}))$$
$$\llbracket \Gamma \vdash \mathsf{Y}_\sigma(M) \rrbracket(\vec{d}) = \mu(\llbracket \Gamma \vdash M \rrbracket(\vec{d}))$$
$$\llbracket \Gamma \vdash \mathsf{zero} \rrbracket(\vec{d}) = 0$$
$$\llbracket \Gamma \vdash \mathsf{succ}(M) \rrbracket(\vec{d}) = \llbracket \mathsf{succ} \rrbracket(\llbracket \Gamma \vdash M \rrbracket(\vec{d}))$$

$$\llbracket \Gamma \vdash \mathsf{pred}(M) \rrbracket (\vec{d}) = \llbracket \mathsf{pred} \rrbracket (\llbracket \Gamma \vdash M \rrbracket (\vec{d}))$$

$$\llbracket \Gamma \vdash \mathsf{ifz}(M, N_1, N_2) \rrbracket (\vec{d}) = \llbracket \mathsf{ifz} \rrbracket (\llbracket \Gamma \vdash M \rrbracket (\vec{d}), \llbracket \Gamma \vdash N_1 \rrbracket (\vec{d}), \llbracket \Gamma \vdash N_2 \rrbracket (\vec{d}))$$

where $\llbracket \mathsf{succ} \rrbracket, \llbracket \mathsf{pred} \rrbracket : N \to N$ are defined as

$$\llbracket \mathsf{succ} \rrbracket (\bot) = \bot \qquad \llbracket \mathsf{succ} \rrbracket (n) = n{+}1$$

$$\llbracket \mathsf{pred} \rrbracket (\bot) = \bot \qquad \llbracket \mathsf{pred} \rrbracket (0) = 0 \qquad \llbracket \mathsf{pred} \rrbracket (n{+}1) = n$$

and $\llbracket \mathsf{ifz} \rrbracket : N^3 \to N$ is defined as

$$\llbracket \mathsf{ifz} \rrbracket (\bot, x, y) = \bot \qquad \llbracket \mathsf{ifz} \rrbracket (0, x, y) = x \qquad \llbracket \mathsf{ifz} \rrbracket (n{+}1, x, y) = y$$

with n ranging over $\mathbb{N}$ and x and y ranging over $N = \mathbb{N}_\bot$. $\diamond$

Notice that $\llbracket \mathsf{succ} \rrbracket$ and $\llbracket \mathsf{pred} \rrbracket$ are the strict extensions of the ordinary successor and predecessor functions on $\mathbb{N}$. The function $\llbracket \mathsf{ifz} \rrbracket$ is strict in its first argument but non-strict in its second and third argument.

We now collect a few basic properties of the domain interpretation of PCF whose proof is fairly standard.

Lemma 3.15   (Substitution Lemma)
Let $\Gamma \equiv x_1{:}\sigma_1, \ldots, x_n{:}\sigma_n$ be a context and $\Gamma \vdash M : \tau$ a term. Then for all contexts Δ and terms $\Delta \vdash N_i : \sigma_i$ with $i{=}1, \ldots, n$ it holds that

$$\llbracket \Delta \vdash M[\vec{N}/\vec{x}] \rrbracket (\vec{d}) = \llbracket \Gamma \vdash M \rrbracket (\llbracket \Delta \vdash N_1 \rrbracket (\vec{d}), \ldots, \llbracket \Delta \vdash N_n \rrbracket (\vec{d}))$$

for all $\vec{d} \in \llbracket \Delta \rrbracket$.

Proof.   Straightforward by induction on derivations of $\Gamma \vdash M$. $\square$

As corollaries to the Substitution Lemma we obtain correctness of the β- and η-equalities as known from λ-calculus.

Corollary 3.16   (β-equality)
If $\Gamma, x{:}\sigma \vdash M : \tau$ and $\Gamma \vdash N : \sigma$ then

$$\llbracket \Gamma \vdash (\lambda x{:}\sigma.\, M)(N) \rrbracket = \llbracket \Gamma \vdash M[N/x] \rrbracket \ .$$

Proof.   Straightforward exercise! $\square$

Corollary 3.17   (η-equality)
If $\Gamma \vdash M : \sigma{\to}\tau$ then

$$\llbracket \Gamma \vdash \lambda x{:}\sigma.\, M(x) \rrbracket = \llbracket \Gamma \vdash M \rrbracket$$

for $x \notin \mathrm{Var}(\Gamma)$.

Proof.   Straightforward exercise!   □

The next theorem relates the domain–theoretic fixpoint semantics to our operational intuition about recursive definitions.

Theorem 3.18   *Let $\Omega_\sigma \equiv Y_\sigma(\lambda x{:}\sigma.\,x)$. Obviously, we have $[\![\Omega_\sigma]\!] = \bot$. For every term $\Gamma \vdash M : \sigma{\to}\sigma$ we have*

$$[\![\Gamma \vdash Y_\sigma(M)]\!] = \bigsqcup_{n \in \mathbb{N}} [\![\Gamma \vdash M^n(\Omega_\sigma)]\!]$$

where $M^n(\Omega_\sigma)$ is defined recursively as $M^0(\Omega_\sigma) \equiv \Omega_\sigma$ and $M^{n+1}(\Omega_\sigma) \equiv M(M^n(\Omega_\sigma))$.

Proof.   As the interpretation of $\lambda x{:}\sigma.\,x$ is id_{D_σ} whose least fixpoint is $\bot$ we get $[\![\Omega_\sigma]\!] = \bot$. Now if $\Gamma \vdash M : \sigma{\to}\sigma$ then for all $\vec{d} \in [\![\Gamma]\!]$ we have

$$[\![\Gamma \vdash Y_\sigma(M)]\!](\vec{d}) = \mu([\![\Gamma \vdash M]\!](\vec{d})) = \bigsqcup_{n \in \mathbb{N}} ([\![\Gamma \vdash M]\!](\vec{d}))^n(\bot)$$

from which it follows that $[\![\Gamma \vdash Y_\sigma(M)]\!] = \bigsqcup_{n \in \mathbb{N}}[\![\Gamma \vdash M^n(\Omega_\sigma)]\!]$ because a straightforward inductive argument shows that

$$[\![\Gamma \vdash M^n(\Omega_\sigma)]\!](\vec{d}) = ([\![\Gamma \vdash M]\!](\vec{d}))^n(\bot)$$

holds for all $n \in \mathbb{N}$ and $\vec{d} \in [\![\Gamma]\!]$.   □

This theorem guarantees that the meaning of a recursive definition $Y_\sigma(M)$ is the supremum of its finite unfoldings $M^n(\Omega_\sigma)$.

3.3   LCF – A Logic of Computable Functionals

At the same time (end of 1960ies) when Dana Scott introduced the prototypical programming language PCF in [Scott 1969] he also defined LCF (Logic of Computable Functionals) for the purpose of reasoning about PCF programs. Due to Gödel's Incompleteness Theorem such a formal system[2] can never be complete since it contains arithmetic. Despite its inherent incompleteness LCF contains a lot of most useful reasoning principles which we will now discuss informally.

Atomic propositions about PCF terms have the form $M \sqsubseteq_\sigma N$ where both M and N are terms of type σ. The relations $\sqsubseteq_\sigma$ are all supposed as

---

[2]Here "formal system" means that its set of theorems is recursively enumerable!

reflexive and transitive. Equality for terms of type σ is then defined as

$$M =_\sigma N \quad \equiv \quad (M \sqsubseteq_\sigma N) \wedge (N \sqsubseteq_\sigma M) \ .$$

The following axioms are easily validated w.r.t. the Scott model introduced in the previous section. Notice, however, that we always assume free variables to be typed by a context which we prefer to leave implicit[3] for sake of readability.

(1) $M_1 \sqsubseteq_{\sigma \to \tau} M_2 \wedge N_1 \sqsubseteq_\sigma N_2 \Rightarrow M_1(N_1) \sqsubseteq_\tau M_2(N_2)$

(2) $\lambda x{:}\sigma.M_1 \sqsubseteq_{\sigma \to \tau} \lambda x{:}\sigma.M_2 \Leftrightarrow \forall x{:}\sigma.\, M_1 \sqsubseteq_\tau M_2$

(3) $(\lambda x{:}\sigma.M)(N) =_\tau M[N/x]$

(4) $\lambda x{:}\sigma.M(x) =_{\sigma \to \tau} M \quad$ provided x is not free in M

(5) $\mathsf{Y}_\sigma(M) =_\sigma M(\mathsf{Y}_\sigma(M))$

(6) $\forall x{:}\sigma.\, M(x) \sqsubseteq_\sigma x \Rightarrow \mathsf{Y}_\sigma(M) \sqsubseteq_\sigma x \quad$ provided x is not free in M

(7) $P(\Omega_\sigma) \wedge (\forall x{:}\sigma.\, P(x) \Rightarrow P(M(x))) \Rightarrow P(\mathsf{Y}_\sigma(M))$

provided x is not free in M and $P(x)$ is a predicate built from atomic formulas via $\forall, \wedge, \vee$ and $A \Rightarrow (-)$ where A is an arbitrary formula without free occurrences of the variable x.

Notice that the syntactic restrictions on the predicate P in (7) guarantee that the predicate P is admissible.[4] Whereas (7) corresponds to Fixpoint Induction (6) corresponds to Park Induction from which one may derive that $\Omega_\sigma = \mathsf{Y}_\sigma(\lambda x{:}\sigma.x) \sqsubseteq_\sigma M$ for all terms M of type σ.

But we also need some axioms for the data type of natural numbers. For their formulation we employ the predicate

$$N(x) \quad \equiv \quad \neg(x = \Omega_{\mathbf{nat}})$$

expressing that x is a proper natural number different from $\bot$. The following axioms are (essentially) those of Peano Arithmetic

(8) $\neg\, \mathsf{succ}(x) = \mathsf{zero}$

(9) $\forall x, y{:}\mathbf{nat}.\ N(x) \wedge N(y) \wedge \mathsf{succ}(x) = \mathsf{succ}(y) \Rightarrow x = y$

---

[3]That means that "officially" one would have to consider instead of formulas A rather expressions of the form $\Gamma \vdash A$ where all free variables of A are declared in Γ.

[4]All these closure properties are straightforward to verify with the exception of disjunction. For this purpose suppose P and Q are admissible subsets of a cpo A. Suppose X is a directed subset of A with $X \subseteq P \cup Q$. If $P \cap X$ is cofinal in X, i.e. for all $x \in X$ there exists $y \in P \cap X$ with $x \sqsubseteq y$ then $P \cap X$ is directed and $\bigsqcup X = \bigsqcup P \cap X \in P$ from which it follows that $\bigsqcup X = \bigsqcup P \cap X \in P$ since P is admissible. Otherwise, i.e. if $P \cap X$ is not cofinal in X, there exists an $x \in X$ with $x \not\sqsubseteq y$ for all $y \in P \cap X$, i.e. for all $y \in X$ with $x \sqsubseteq y$ it holds that $y \notin P$ and thus $y \in Q$, from which it follows that $Q \cap X$ is directed and has the same supremum as X and thus $\bigsqcup X = \bigsqcup Q \cap X \in Q$ since Q is admissible.

(10) $P(\mathsf{zero}) \wedge \big(\forall x{:}\mathbf{nat}.\, N(x) \wedge P(x) \Rightarrow P(\mathsf{succ}(x))\big) \Rightarrow \forall x{:}\mathbf{nat}.\, N(x) \Rightarrow P(x)$

where P in (10) is an arbitrary predicate. Finally we need a few axioms governing the use of the basic functions

(11) $N(\mathsf{zero})$

(12) $\forall x{:}\mathbf{nat}.\, N(x) \Leftrightarrow N(\mathsf{succ}(x))$

(13) $\mathsf{pred}(\mathsf{zero}) = \mathsf{zero}$

(14) $\forall x{:}\mathbf{nat}.\, \mathsf{pred}(\mathsf{succ}(x)) = x$

(15) $\mathsf{ifz}(\Omega_{\mathbf{nat}}, x, y) = \Omega_{\mathbf{nat}}$

(16) $\mathsf{ifz}(\mathsf{zero}, x, y) = x$

(17) $\forall z{:}\mathbf{nat}.\, N(z) \Rightarrow \mathsf{ifz}(\mathsf{succ}(z), x, y) = y.$

These axioms turn out as sufficient in practice for verifying properties of PCF programs. We have already mentioned that for principal reasons there cannot be a complete axiomatization of the Scott model. However, it is an interesting open problem to find a set of axioms for PCF which together with all true sentences of arithmetic allow one to derive all sentences true in the Scott model. The list of axioms given above surely does not have this property as they are known to be valid in models of PCF different from the Scott model as e.g. the fully abstract model of PCF.

Chapter 4

Computational Adequacy

One easily verifies by induction on the structure of derivations that in the Scott model we have $[\![M]\!] = [\![V]\!]$ whenever $M{\Downarrow}V$. Thus, the Scott model is *correct* w.r.t. the operational semantics. This holds in particular for programs, i.e. closed terms of type **nat**. In this chapter we will show that the Scott model is also *computationally adequate* for the operational semantics, i.e. that $M{\Downarrow}\underline{n}$ whenever $[\![M]\!] = n \in \mathbb{N}$.[1]

At first one might think of proving computational adequacy by induction on the structure of programs. This, however, is impossible because subterms of programs, i.e. closed terms of type **nat**, need neither be closed nor of type **nat**. The first problem is not that serious as one may quantify over all closed instances of open terms. The second problem, however, is much more serious and requires the introduction of a new concept, namely that of a *logical relation*. This notion and its variations will turn out as a key concept for semantic investigations since—as we shall see later on—it has much more applications than just providing a nice perspicuous proof of computational adequacy.

Our proof of computational adequacy will be organised as follows. We define for each type σ a relation

$$R_\sigma \subseteq D_\sigma \times \mathsf{Prg}_\sigma$$

by induction on the structure of σ. The family $R = (R_\sigma \,|\, \sigma \in \mathsf{Type})$ will be called a *logical relation* (between semantics and syntax of PCF).

---

[1]Notice that for types different from **nat** one cannot expect that $M{\Downarrow}V$ whenever $[\![M]\!] = [\![V]\!]$ as is immediate from considering the different syntactic values

$$M \equiv \lambda x{:}\mathbf{nat}.x \qquad \text{and} \qquad V \equiv \lambda x{:}\mathbf{nat}.\mathsf{pred}(\mathsf{succ}(x))$$

whose interpretation is equal, namely the identity function on $\mathbb{N}_\perp$.

Then we prove that $[\![M]\!]R_\sigma M$ holds for all closed terms M of type σ. As the logical relation will be defined in a way that $[\![M]\!]R_{\mathbf{nat}}M$ just means $\forall n{\in}\mathbb{N}.\,[\![M]\!]{=}n \Rightarrow M{\Downarrow}\underline{n}$ we are done.

Now we define the logical relation needed for our proof of computational adequacy.

Definition 4.1   We define a family $R = (R_\sigma \,|\, \sigma \in \mathsf{Type})$ of relations $R_\sigma \subseteq D_\sigma \times \mathsf{Prg}_\sigma$ via the clauses

$$dR_{\mathbf{nat}}M \quad \text{iff} \quad \forall n{\in}\mathbb{N}.\,d{=}n \Rightarrow M{\Downarrow}\underline{n}$$

$$fR_{\sigma\to\tau}M \quad \text{iff} \quad \forall d{\in}D_\sigma.\forall N{\in}\mathsf{Prg}_\sigma.\,dR_\sigma N \Rightarrow f(d)R_\tau M(N)$$

by induction on the structure of types.   ◇

Notice that for $\sigma \equiv \sigma_1{\to}\ldots{\to}\sigma_k{\to}\mathbf{nat}$ we can reformulate $fR_\sigma M$ as

$$\forall d_1 R_{\sigma_1} N_1.\ldots.\forall d_k R_{\sigma_k} N_k.\,f(d_1)\ldots(d_k)R_{\mathbf{nat}}M(N_1)\ldots(N_k)$$

or, more explicitly, as

$$\forall d_1 R_{\sigma_1} N_1.\ldots.\forall d_k R_{\sigma_k} N_k.\,\forall n{\in}\mathbb{N}.\,f(d_1)\ldots(d_k){=}n \Rightarrow M(N_1)\ldots(N_k){\Downarrow}\underline{n}\;.$$

In any case it is obvious from the definition of $R_{\mathbf{nat}}$ that computational adequacy is equivalent to $[\![M]\!]R_{\mathbf{nat}}M$ for all programs M of type $\mathbf{nat}$.

For the purpose of showing that $[\![M]\!]R_\sigma M$ holds for all $M \in \mathsf{Prg}_\sigma$ we need some properties of the logical relation R.

Lemma 4.2   *For all types σ it holds that*

(1) *if $d' \sqsubseteq d$ and $dR_\sigma M$ then $d'R_\sigma M$*
(2) *for every $M \in \mathsf{Prg}_\sigma$ the set $R_\sigma M := \{\,d \in D_\sigma \,|\, dR_\sigma M\,\}$ is closed under directed suprema and contains $\bot$*
(3) *if $dR_\sigma M$ and $M \lesssim_\sigma M'$ then $dR_\sigma M'$.*

Proof.   Obviously, conditions (1)–(3) hold for base type $\mathbf{nat}$. For the general case $\sigma \equiv \sigma_1{\to}\ldots{\to}\sigma_k{\to}\mathbf{nat}$ we employ the characterisation of R_σ as given in the remark after Definition 4.1.

ad (1) :   Suppose $g \sqsubseteq f$ and $fR_\sigma M$.   For showing $gR_\sigma M$ suppose that $d_i R_{\sigma_i} N_i$ for $i{=}1,\ldots,k$.   From $fR_\sigma M$ it follows that $f(d_1)\ldots(d_k)R_{\mathbf{nat}}M(N_1)\ldots(N_k)$.   Since $g \sqsubseteq f$ we have $g(d_1)\ldots(d_k) \sqsubseteq f(d_1)\ldots(d_k)$.   Thus, since (1) holds for base type it follows that $g(d_1)\ldots(d_k)R_{\mathbf{nat}}M(N_1)\ldots(N_k)$ as desired.

ad (2) : Obviously, we have $\perp R_\sigma M$ since (2) holds for base type and $\perp(d_1)\ldots(d_k) = \perp$ for all $d_i \in D_{\sigma_i}$. For closure under directed suprema suppose that $F \subseteq \{d \in D_\sigma \mid dR_\sigma M\}$ is directed. For showing that $\bigsqcup F R_\sigma M$ suppose that $d_i R_{\sigma_i} N_i$ for $i=1,\ldots,k$. Then for all $f \in F$ we have $f(d_1)\ldots(d_k)R_{\mathbf{nat}}M(N_1)\ldots(N_k)$ since $fR_\sigma M$. As $(\bigsqcup F)(d_1)\ldots(d_k) = \bigsqcup_{f\in F} f(d_1)\ldots(d_k)$ and (2) holds for base type it follows that $(\bigsqcup F)(d_1)\ldots(d_k)R_{\mathbf{nat}}M(N_1)\ldots(N_k)$ as desired.

ad (3) : Suppose $fR_\sigma M$ and $M \sqsubseteq_\sigma M'$. For showing $fR_\sigma M'$ suppose that $d_i R_{\sigma_i} N_i$ for $i=1,\ldots,k$. Since $fR_\sigma M$ we have $f(d_1)\ldots(d_k)R_{\mathbf{nat}}M(N_1)\ldots(N_k)$. From $M \sqsubseteq_\sigma M'$ it follows that $M(N_1)\ldots(N_k) \sqsubseteq_{\mathbf{nat}} M'(N_1)\ldots(N_k)$. Thus, since (3) holds for base type it follows that $f(d_1)\ldots(d_k)R_{\mathbf{nat}}M'(N_1)\ldots(N_k)$ as desired. $\qquad\square$

We also need that the least fixpoint operator on D_σ is related to Y_σ via $R_{(\sigma\to\sigma)\to\sigma}$. For showing this we need the following lemma.

Lemma 4.3 *For every PCF type σ it holds that $M(\mathsf{Y}_\sigma(M)) \sqsubseteq_\sigma \mathsf{Y}_\sigma(M)$.*

Proof. Suppose that $M(\mathsf{Y}_\sigma(M))(N_1)\ldots(N_k)\Downarrow\underline{n}$. Then by inspection of the inductive definition of $\Downarrow$ there exist syntactic values $V_1,\ldots,V_k$ such that $M(\mathsf{Y}_\sigma(M))\Downarrow V_1$, $V_i(N_i)\Downarrow V_{i+1}$ for $i < k$ and $V_k(N_k)\Downarrow\underline{n}$. But from $M(\mathsf{Y}_\sigma(M))\Downarrow V_1$ it follows that $\mathsf{Y}_\sigma(M)\Downarrow V_1$ and, accordingly, that $\mathsf{Y}_\sigma(M)(N_1)\ldots(N_k)\Downarrow\underline{n}$ as desired. $\qquad\square$

A similar argument shows that $\mathsf{Y}_\sigma(M) \sqsubseteq_\sigma M(\mathsf{Y}_\sigma(M))$. Thus, by Milner's Context Lemma (proved in the next chapter) the terms $\mathsf{Y}_\sigma(M)$ and $M(\mathsf{Y}_\sigma(M))$ are observationally equivalent as expected.

Lemma 4.4 *If $fR_{\sigma\to\sigma}M$ then $\mu(f)R_\sigma\mathsf{Y}_\sigma(M)$.*

Proof. Suppose $fR_{\sigma\to\sigma}M$. For showing $\mu(f)R_\sigma\mathsf{Y}_\sigma(M)$ by Lemma 4.2(2) it suffices to show that $f^n(\perp)R_\sigma\mathsf{Y}_\sigma(M)$ for all $n \in \mathbb{N}$. The base case $\perp R_\sigma\mathsf{Y}_\sigma(M)$ holds by Lemma 4.2(2). Suppose $f^n(\perp)R_\sigma\mathsf{Y}_\sigma(M)$ as induction hypothesis. Thus, as $fR_\sigma M$ it follows that $f^{n+1}(\perp)R_\sigma M(\mathsf{Y}_\sigma(M))$. By Lemma 4.3 we have $M(\mathsf{Y}_\sigma(M)) \sqsubseteq_\sigma \mathsf{Y}_\sigma(M)$. Using Lemma 4.2(3) conclude that $f^{n+1}(\perp)R_\sigma\mathsf{Y}_\sigma(M)$ as desired. $\qquad\square$

Now we are ready to prove the following *Main Lemma* for the logical relation R entailing that $[\![M]\!]R_\sigma M$ for all $M \in \mathsf{Prg}_\sigma$.

Lemma 4.5 (Main Lemma for R)
If $x_1{:}\sigma_1,\ldots,x_k{:}\sigma_k \vdash M : \tau$ and $d_i R_{\sigma_i} N_i$ for $i=1,\ldots,k$ then

$$[\![x_1{:}\sigma_1,\ldots,x_k{:}\sigma_k \vdash M]\!](\vec{d})\; R_\tau\; M[\vec{N}/\vec{x}] \;.$$

Proof.   The proof is by induction on the structure of derivations of terms in context. For facilitating notation we write $\vec{d}R\vec{N}$ instead of $d_i R_{\sigma_i} N_i$ for $i=1,\ldots,k$.

Variables : For $x_1{:}\sigma_1,\ldots,x_k{:}\sigma_k \vdash x_i : \sigma_i$ and $\vec{d}R\vec{N}$ we have

$$[\![x_1{:}\sigma_1,\ldots,x_k{:}\sigma_k \vdash x_i]\!](\vec{d}) = d_i\, R_{\sigma_i}\, N_i \equiv x_i[\vec{N}/\vec{x}]$$

as desired.

λ-Abstraction : Suppose as induction hypothesis that the claim of the theorem holds for $\Gamma, x{:}\sigma \vdash M : \tau$. Further suppose that $\vec{d}R\vec{N}$. We have to show that

$$[\![\Gamma \vdash \lambda x{:}\sigma.\, M]\!](\vec{d})\, R_{\sigma\to\tau}\, (\lambda x{:}\sigma.\, M)[\vec{N}/\vec{x}]$$

where $\vec{x}$ is the list of variables declared in Γ. For that purpose assume that $dR_\sigma N$. From the induction hypothesis it follows that

$$[\![\Gamma, x{:}\sigma \vdash M]\!](\vec{d},d)\, R_\tau\, M[\vec{N},N/\vec{x},x]$$

and, therefore, by Lemma 4.2(3) that

$$[\![\Gamma \vdash \lambda x{:}\sigma.\, M]\!](\vec{d})(d)\, R_\tau\, (\lambda x{:}\sigma.\, M)[\vec{N}/\vec{x}](N)$$

because

$$[\![\Gamma \vdash \lambda x{:}\sigma.\, M]\!](\vec{d})(d) = [\![\Gamma, x{:}\sigma \vdash M]\!](\vec{d},d)$$

and[2]

$$M[\vec{N},N/\vec{x},x] \equiv M[\vec{N}/\vec{x}][N/x] \,\gtrsim_\tau\, (\lambda x{:}\sigma.\, M[\vec{N}/\vec{x}])(N) \equiv (\lambda x{:}\sigma.\, M)[\vec{N}/\vec{x}](N)\ .$$

Application : Suppose as induction hypothesis that the claim of the theorem holds for $\Gamma \vdash M_1 : \sigma\to\tau$ and $\Gamma \vdash M_2 : \sigma$. Now if $\vec{d}R\vec{N}$ then due to the induction hypotheses we have

$$[\![\Gamma \vdash M_i]\!](\vec{d})\, R\, M_i[\vec{N}/\vec{x}]$$

for $i=1,2$ from which it follows that

$$[\![\Gamma\vdash M_1]\!](\vec{d})([\![\Gamma \vdash M_2]\!](\vec{d}))\, R\, M_1[\vec{N}/\vec{x}](M_2[\vec{N}/\vec{x}])\ .$$

As

$$[\![\Gamma\vdash M_1(M_2)]\!](\vec{d}) = [\![\Gamma\vdash M_1]\!](\vec{d})([\![\Gamma \vdash M_2]\!](\vec{d}))$$

---

[2]as it generally holds that $M[N/x] \gtrsim (\lambda x{:}\sigma.M)N$ for $M \in \mathsf{Prg}_{\sigma\to\tau}$ and $N \in \mathsf{Prg}_\sigma$

and

$$M_1[\vec{N}/\vec{x}](M_2[\vec{N}/\vec{x}]) \equiv M_1(M_2)[\vec{N}/\vec{x}]$$

it then follows that

$$[\![\Gamma \vdash M_1(M_2)]\!](\vec{d}) \; R \; M_1(M_2)[\vec{N}/\vec{x}]$$

as desired.

Recursion : Suppose as induction hypothesis that $\Gamma \vdash M : \sigma \rightarrow \sigma$ satisfies the requirement of the theorem. Now if $\vec{d}R\vec{N}$ then due to the induction hypothesis we have

$$[\![\Gamma \vdash M]\!](\vec{d}) \; R \; M[\vec{N}/\vec{x}]$$

from which it follows by Lemma 4.4 that

$$[\![\mathsf{Y}_\sigma(M)]\!](\vec{d}) = \mu([\![\Gamma \vdash M]\!](\vec{d})) \; R \; \mathsf{Y}_\sigma(M[\vec{N}/\vec{x}]) \equiv \mathsf{Y}_\sigma(M)[\vec{N}/\vec{x}]$$

as desired.

Basic Operations : One easily checks that $0R_{\mathbf{nat}}\mathsf{zero}$ and that from $xR_{\mathbf{nat}}M$, $y_1R_{\mathbf{nat}}N_1$ and $y_2R_{\mathbf{nat}}N_2$ it follows that $[\![\mathsf{succ}]\!](x)R_{\mathbf{nat}}\mathsf{succ}(M)$, $[\![\mathsf{pred}]\!](x)R_{\mathbf{nat}}\mathsf{pred}(M)$ and $[\![\mathsf{ifz}]\!](x,y_1,y_2)R_{\mathbf{nat}}\mathsf{ifz}(M,N_1,N_2)$. Using these observations the remaining cases for zero, succ, pred and ifz go through without pain. □

The following theorem is a special case of the previous lemma.

Theorem 4.6 (Computational Adequacy)
For every closed term M of type **nat** *we have* $M \Downarrow \underline{n}$ *whenever* $[\![M]\!] = n$.

Proof. Immediate from Lemma 4.5 since $\vdash M : \mathbf{nat}$ for closed terms M of type **nat**. □

Thus, for $\vdash M : \mathbf{nat}$ and $n \in \mathbb{N}$ we have

$$M \Downarrow \underline{n} \qquad \text{iff} \qquad [\![M]\!] = n$$

i.e. a closed term of type **nat** denotes a natural number if and only if it evaluates to the corresponding numeral.

Chapter 5

Milner's Context Lemma

We now will use the logical relation R of the previous chapter to cook up a slick proof of Milner's Context Lemma saying that $\sqsubseteq$ and $\lesssim$ coincide at all types. Whereas older proofs were fairly syntactical and combinatorial in character the current proof (due to A. Jung) is fairly abstract and, accordingly, more transparent.

Theorem 5.1 (Milner's Context Lemma)
For all types σ and $M, N \in \mathsf{Prg}_\sigma$ the following conditions are equivalent

   (a) $M \sqsubseteq_\sigma N$ (b) $M \lesssim_\sigma N$ (c) $[\![M]\!] R_\sigma N$.

Proof.   (b) $\Rightarrow$ (a) : since contexts of the form $[\,]\vec{P}$ are just particular contexts of base type.
(a) $\Rightarrow$ (c) : Suppose $M \sqsubseteq_\sigma N$. From Lemma 4.5 we know that $[\![M]\!] R M$. Thus, it follows from Lemma 4.2(3) that $[\![M]\!] R_\sigma N$.
(c) $\Rightarrow$ (b) : Suppose $[\![M]\!] R_\sigma N$. Let $P \in \mathsf{Prg}_{\sigma \to \mathsf{nat}}$. As by Lemma 4.5 we have $[\![P]\!] R_{\sigma \to \mathsf{nat}} P$ it follows that $[\![P]\!]([\![M]\!]) R_\mathsf{nat} N(P)$. Thus, as $[\![P(M)]\!] = [\![P]\!]([\![M]\!])$ we have $[\![P(M)]\!] R_\mathsf{nat} P(N)$. If $P(M){\Downarrow}\underline{n}$ then by correctness of the operational semantics we have $[\![M(P)]\!] = n$ and, therefore, also $n R_\mathsf{nat} P(N)$ from which it follows that $P(N){\Downarrow}\underline{n}$ as desired. $\qquad\square$

The following corollary gives a further characterisation of $\lesssim$.

Corollary 5.2   *For all types σ and $M, N \in \mathsf{Prg}_\sigma$ we have*

$$M \lesssim_\sigma N \qquad \textit{iff} \qquad \forall d{\in}D_\sigma.\, d R_\sigma M \Rightarrow d R_\sigma N \ .$$

Proof.   The forward direction follows from Lemma 4.2(3) since by Theorem 5.1 the relations $\lesssim_\sigma$ and $\sqsubseteq_\sigma$ coincide.
For the reverse direction suppose that $\forall d{\in}D_\sigma.\, d R_\sigma M \Rightarrow d R_\sigma N$. Then, in particular, we have $[\![M]\!] R_\sigma M \Rightarrow [\![M]\!] R_\sigma N$. As by Lemma 4.5 we have

$[\![M]\!]R_\sigma M$ it follows that $[\![M]\!]R_\sigma N$. Thus, by Theorem 5.1 it follows that $M \lesssim_\sigma N$ as desired.   □

This corollary tells us that we may replace quantification over syntactic experiments of the form $P(-){\Downarrow}\underline{n}$ equivalently by quantification over semantic experiments of the form $dR_\sigma(-)$.

We may also use the logical relation R for defining an alternative partial order $\leq_\sigma$ on D_σ as follows

$$d_1 \leq_\sigma d_2 \qquad \text{iff} \qquad \forall P \in \mathsf{Prg}_\sigma.\ d_2 R_\sigma P \Rightarrow d_1 R_\sigma P$$

which by Lemma 4.2(1) contains $\sqsubseteq_{D_\sigma}$. It is an easy exercise(!) to show that $\leq_\sigma$ (as a subset of $D_\sigma \times D_\sigma$) is closed under suprema of directed sets. For closed terms M and N of type σ we have

$$[\![M]\!] \leq_\sigma [\![N]\!] \qquad \text{iff} \qquad M \lesssim_\sigma N$$

as by Theorem 5.1 the condition $\forall P \in \mathsf{Prg}_\sigma.\ [\![N]\!]R_\sigma P \Rightarrow [\![M]\!]R_\sigma P$ is equivalent to $\forall P \in \mathsf{Prg}_\sigma.\ N \lesssim_\sigma P \Rightarrow M \lesssim_\sigma P$, i.e. $M \lesssim_\sigma N$. Thus, in a very precise sense $\leq$ is the denotational analogue of $\lesssim$.

Chapter 6

The Full Abstraction Problem

In the previous chapter we have seen that $[\![M]\!] \sqsubseteq [\![N]\!]$ entails $M \lesssim N$ as from $[\![M]\!] \sqsubseteq [\![N]\!]$ and $[\![N]\!]RN$ it follows that $[\![M]\!]RN$ and thus $M \lesssim N$. In this chapter we will show that the reverse implication is not valid for the Scott model. Even more we will show that $[\![M]\!] = [\![N]\!]$ need not hold even if M and N are observationally equal. This phenomenon occurs already at type $(\iota{\rightarrow}\iota{\rightarrow}\iota){\rightarrow}\iota$ and the reason is that there are *not enough* PCF-*definable objects* within $D_{\iota\rightarrow\iota\rightarrow\iota}$.

For the reader's convenience we officially fix some terminology introduced informally already in the introduction.

Definition 6.1 (Full Abstraction)
A model of PCF is called *equationally fully abstract* iff

$$M \simeq N \Rightarrow [\![M]\!] = [\![N]\!]$$

for all closed terms M and N of the same type where $M \simeq N$ is an abbreviation for $M \lesssim N \wedge N \lesssim N$.
A model of PCF is called *fully abstract* iff

$$M \lesssim N \Rightarrow [\![M]\!] \sqsubseteq [\![N]\!]$$

for all closed terms M and N of the same type.   ◊

Obviously, every fully abstract model is also equationally fully abstract.

In the next chapter we will show that there is no PCF-definable function $f \in D_{\iota\rightarrow\iota\rightarrow\iota}$ satisfying the constraints

$$(\dagger) \qquad f0\bot = 0 \qquad f\bot0 = 0 \qquad f11 = 1 \ .$$

In the Scott model there is a least one such function, namely

$$\text{por}xy = \begin{cases} 0 & \text{if } x = 0 \text{ or } y = 0 \\ 1 & \text{if } x = 1 = y \\ \bot & \text{otherwise} \end{cases}$$

called "parallel or". In the following lemma we exhibit two functionals of type $(\iota \to \iota \to \iota) \to \iota$ which are definable in PCF and give different results only when applied to f satisfying condition (†).

Lemma 6.2 *Consider the terms*

$$\text{portest}_i \equiv \lambda f{:}\iota \to \iota \to \iota.\ \text{ifz}(f\,\underline{0}\,\Omega_\iota, \text{ifz}(f\,\Omega_\iota\,\underline{0}, \text{ifz}(\widetilde{\text{pred}}(f\,\underline{1}\,\underline{1}), \underline{i}, \Omega_\iota), \Omega_\iota), \Omega_\iota)$$

of type $(\iota \to \iota \to \iota) \to \iota$ for $i = 0, 1$ where $\Omega_\iota \equiv Y_\iota(\lambda x{:}\iota.x)$ and $\widetilde{\text{pred}}$ stands for the term $\lambda x{:}\iota.\text{ifz}(x, \Omega_\iota, \text{ifz}(\text{pred}(x), \underline{0}, \Omega_\iota))$. Then for all $f \in D_{\iota \to \iota \to \iota}$

- $[\![\text{portest}_i]\!](f) = i$ *whenever f satisfies condition (†) and*
- $[\![\text{portest}_i]\!] = \bot$ *otherwise.*

Proof. Obvious by unfolding the definition of portest_0 and portest_1. □

Thus, if for all closed terms of type $\iota \to \iota \to \iota$ their interpretation does not satify condition (†) then by Milner's Context Lemma (and computational adequacy) the programs $\text{portest}_0'$ and portest_1 are observationally equal although their interpretations are different in the Scott model since $[\![\text{portest}_0]\!](\text{por}) = 0$ and $[\![\text{portest}_1]\!](\text{por}) = 1$.

Intuitively, it is clear that one cannot implement a function satisfying (†) in PCF because evaluation strategies for PCF terms are necessarily sequential, i.e. either the first or the second argument has to be evaluated first, whereas any implementation of an f satisfying condition (†) has to evaluate both arguments *in parallel*.[1] However, a precise mathematical proof of this fact requires some sophistication and, therefore, we postpone it to the next chapter where we develop the necessary machinery of logical relations.

Actually, we will show a bit more, namely that all first order PCF-definable functions are *stable*, i.e. preserve binary infima (denoted by $\sqcap$) of consistent pairs, i.e. pairs of elements having a common extension[2].

---

[1] For example the first argument may diverge whereas the second argument evaluates to 0. In this case it would be wrong to evaluate the first argument first as this would lead to non-termination of the function call although it should evaluate to 0. A symmetric argument shows that it is also wrong to evaluate always the second argument first.

[2] We write $x \uparrow y$ as an abbreviation for $\exists z.\ x \sqsubseteq z \wedge y \sqsubseteq z$ and say that "x and y are consistent" if this condition holds.

Lemma 6.3 *For every term M of first order type, i.e. of type*

$$\overbrace{\iota \to \ldots \to \iota \to}^{k \text{ times}} \iota$$

for some $k \in \mathbb{N}$, it holds that

$$[\![M]\!](x_1 \sqcap y_1) \ldots (x_k \sqcap y_k) = [\![M]\!](x_1) \ldots (x_k) \sqcap [\![M]\!](y_1) \ldots (y_k)$$

for all $\vec{x}, \vec{y} \in D_\iota^k$ with $x_i \uparrow y_i$ for $i=1, \ldots, k$.

This has the consequence that

Corollary 6.4 *There are no PCF definable functions of type $D_{\iota \to \iota \to \iota}$ satisfying the constraint (†).*

Proof. Suppose f is PCF definable and satisfies (†). Then $f0\bot = 0 = f\bot 0$ and, therefore, we have

$$f\bot\bot = f(0 \sqcap \bot)(\bot \sqcap 0) = f0\bot \sqcap f\bot 0 = 0 \sqcap 0 = 0$$

by Lemma 6.3. However, by (†) we have $f11 = 1$ and, thus, by monotonicity of f it follows that $0 \sqsubseteq 1$ which is impossible. $\qquad\square$

The observation of Lemma 6.3 was taken as a starting point by G. Berry, who in his Thése d' Etat [Berry 1979] introduced and investigated a category of so-called "stable domains" where **all** morphisms are required not only to be Scott continuous but also *stable* in the sense that

$$x \uparrow y \;\Rightarrow\; f(x \sqcap y) = f(x) \sqcap f(y)$$

for all arguments x and y. The obvious advantage of stable domain theory is that it refutes the existence of maps like "parallel or" which—as we have seen—are responsible for the lack of full abstraction of the Scott model. This, however, is achieved only at the price that the order on function spaces is *not pointwise* anymore.

Notice that in the Scott model even very simple PCF-definable functions are not stable as for example the evaluation function

$$\mathsf{ev} = \lambda f{:}\iota \to \iota.\lambda x{:}\iota.f(x)$$

which can be seen as follows. Consider the functions $f_1 = [\![\lambda x{:}\iota.\mathsf{zero}]\!]$ and $f_2 = [\![\lambda x{:}\iota.\mathsf{ifz}(x, \mathsf{zero}, \mathsf{zero})]\!]$. We have $f_1\bot = 0$ and $f_20 = 0$ and, therefore, also $f_1\bot \sqcap f_20 = 0$ whereas $(f_1 \sqcap f_2)(\bot \sqcap 0) = f_2\bot = \bot$ which is a counterexample to stability of ev as $f_2 \sqsubseteq f_1$ and $\bot \sqsubseteq 0$. Accordingly, in the stable domains model $f_2 \not\sqsubseteq f_1$ as otherwise ev were not even monotonic. But we have $\lambda x{:}\iota.\mathsf{ifz}(x, \mathsf{zero}, \mathsf{zero}) \lesssim \lambda x{:}\iota.\mathsf{zero}$ and, accordingly, Berry's stable domains models is not fully abstract either.

A remarkable consequence of the stability of PCF-definable first order functions f is that whenever $f\vec{x} = n \in \iota$ then there exists a least $\vec{y} \sqsubseteq \vec{x}$ with $f\vec{y} = n$, namely the infimum of all $\vec{z} \sqsubseteq \vec{x}$ with $f\vec{z} = n$. Notice that if there is a sequential evaluation strategy for f then this property is automatic. Thus, there is no sequential algorithm for functions f satisfying condition (†) as then $f0\bot = 0 = f\bot 0$ whereas $f\bot\bot = \bot$, i.e. $(0, \bot)$ and $(\bot, 0)$ are different minimal approximations to $(0, 0)$ giving rise to output 0.

An operational semantics for por for which the Scott model is computationally adequate is given by the rules

$$\frac{M{\Downarrow}\underline{0}}{\mathsf{por}MN{\Downarrow}\underline{0}} \qquad \frac{N{\Downarrow}\underline{0}}{\mathsf{por}MN{\Downarrow}\underline{0}} \qquad \frac{M{\Downarrow}\underline{1} \quad N{\Downarrow}\underline{1}}{\mathsf{por}MN{\Downarrow}\underline{1}}$$

where the first two rules do overlap as there are two different derivations of $\mathsf{por}\underline{00}{\Downarrow}\underline{0}$. Although for every term M there is still at most one V with $M{\Downarrow}V$ there is no sequential evaluation strategy for terms of the form $\mathsf{por}MN$. As if one would always evaluate the first argument first then the evaluation of the whole term may diverge even if the second argument evaluates to $\underline{0}$. A symmetric argument shows that it is also wrong to evaluate always the second argument first.

For reasons of efficiency deterministic parallel language constructs like por are not implemented in actual functional languages and, accordingly, the Scott model is not fully abstract for them. However, it should be emphasized that por *is* computable and thus can be implemented in principle. It corresponds to the well known *dove tailing* technique known from recursion theory (see e.g. [Rogers 1987]) where it is used e.g. for showing that semi-decidable predicates are closed under binary unions. Dove tailing as used in recursion theory is highly intensional as it uses Kleene's T-predicate which amounts to a primitive recursive coding[3] of operational semantics: if $A_i = \{n{\in}\mathbb{N} \mid \exists k.T(e_i, n, k)\}$ then $A_1 \cup A_2$ is the halting set of the algorithm $\mu k.T(e_1, n, k) \vee T(e_2, n, k)$. The language construct por may be considered as an extensional version of dove tailing avoiding any coding of the operational semantics. In recursion theory one uses also infinite variants of dove tailing e.g. for showing that the union of an r.e. set of r.e. sets is r.e. again. This can be implemented in PCF using por via the functional

---

[3]Recall that $T(e, n, k)$ means "k is a code for a terminating computation sequence for the application of program with number e to argument n".

$\text{por}_\infty : (\iota{\to}\iota){\to}\iota$ defined via the recursion equation

$$\text{por}_\infty(p) = \text{por}(p(\text{zero}))(\text{por}_\infty(\lambda x{:}\textbf{nat}.p(\text{succ}(x))))$$

from which one can read off easily an implementing PCF term.

In his famous paper "LCF considered as a programming language" [Plotkin 1977] G. Plotkin has shown that the Scott model is fully abstract for PCF+por. The clue of the proof[4] was to show that for every type σ sufficiently many elements, namely the so-called *compact* (or *finite*) elements, are all definable in PCF + por. These compact elements suffice as every element of D_σ appears as directed supremum of the compact elements approximating it. Thus, if continuous functions are identical on the compact elements then they are identical on all arguments. In the same paper [Plotkin 1977] Plotkin has shown that, however, not all "computable" elements of the Scott model can be denoted by terms of PCF+por.[5] This, however, can be remedied by adding the "continuous existential quantifier" $\exists : (\iota{\to}\iota){\to}\iota$ whose operational semantics is given by the rules

$$\frac{M\underline{n}\Downarrow\underline{0}}{\exists(M)\Downarrow\underline{0}}\ (n \in \mathbb{N}) \qquad \frac{M\Omega_\iota\Downarrow\underline{1}}{\exists(M)\Downarrow\underline{1}}$$

Summarizing we can say that the Scott model is neither fully abstract nor universal for PCF. However, this doesn't diminish the relevance of the Scott model as there are reasonable extensions of PCF for which the Scott model is fully abstract and even universal. Much later in Chapters 11 and 12 we will use a somewhat sophisticated logical relation technique for transforming the Scott model into a fully abstract model for PCF.

---

[4] for details see Chapter 13

[5] An element is called "computable" iff the set of codes of approximating compact elements is recursively enumerable.

Chapter 7

Logical Relations

In this chapter we will discuss logical relations on the Scott model for PCF. These will allow us to express invariance properties of syntactic definability without any reference to syntax and use these to prove in a mathematically precise way that certain elements of the Scott model are not PCF definable.

Definition 7.1 *(Logical Relation)*
Let W be an arbitrary set. A *logical relation of arity* W on the Scott model of PCF is a family

$$R = (R_\sigma \in \mathcal{P}(D_\sigma^W) \mid \sigma \in \text{Type})$$

such that

$$f \in R_{\sigma \to \tau} \iff \forall d \in R_\sigma. \underline{\lambda} i \in W. f(i)(d(i)) \in R_\tau \iff \forall d \in R_\sigma. \text{ev} \circ \langle f, d \rangle \in R_\tau$$

for all types σ and τ. $\diamond$

Notice that a logical relation R of arity W is uniquely determined by R_{nat} and that for *all* subsets $P \subseteq D_{\text{nat}}^W$ there exists a unique logical relation R of arity W with $R_{\text{nat}} = P$.

Theorem 7.2 (Main Lemma for Logical Relations)
Let R be a logical relation of arity W on the Scott model of PCF. Then for λ-terms $x_1{:}\sigma_1, \ldots, x_n{:}\sigma_n \vdash M : \tau$ and $d_j \in R_{\sigma_j}$ for $j = 1, .., n$ it holds that

$$\underline{\lambda} i \in W. [\![x_1{:}\sigma_1, \ldots, x_n{:}\sigma_n \vdash M]\!](\vec{d}(i)) \in R_\tau$$

where $\vec{d}(i) = \langle d_1(i), \ldots, d_n(i) \rangle$ for $i \in W$.

Proof. We proceed by induction on the structure of derivations of judgements of the form $\Gamma \vdash M : \sigma$ using only the rules for variables, λ-abstraction and application.

(1) If M is a variable then the claim is trivial (exercise!).

(2) Suppose as induction hypothesis that the claim holds for $\Gamma, x{:}\sigma \vdash M : \tau$ where $\Gamma \equiv x_1{:}\sigma_1, \ldots, x_n{:}\sigma_n$. Suppose that $d_j \in R_{\sigma_j}$ for $j=1, \ldots, n$. We have to show that

$$\underline{\lambda}i{\in}W. \ [\![\Gamma \vdash \lambda x{:}\sigma.M]\!](\vec{d}(i)) \in R_{\sigma \to \tau} \ .$$

For that purpose assume that $d \in R_\sigma$. Then we have

$$\underline{\lambda}i{\in}W. \ [\![\Gamma \vdash \lambda x{:}\sigma.M]\!](\vec{d}(i))(d(i)) = \underline{\lambda}i{\in}W. \ [\![\Gamma, x{:}\sigma \vdash M]\!](\langle\vec{d}(i), d(i)\rangle)$$

whose right hand side is in R_τ due to the induction hypothesis.

(3) Suppose as induction hypotheses that the claim holds for $\Gamma \vdash M : \sigma{\to}\tau$ and $\Gamma \vdash N : \sigma$ where $\Gamma \equiv x_1{:}\sigma_1, \ldots, x_n{:}\sigma_n$. Let $d_j \in R_{\sigma_j}$ for $j=1, \ldots, n$. We have to show that

$$\underline{\lambda}i{\in}W. \ [\![\Gamma \vdash M(N)]\!](\vec{d}(i)) \in R_\tau \ .$$

As $[\![\Gamma \vdash M(N)]\!](\vec{d}(i)) = [\![\Gamma \vdash M]\!](\vec{d}(i))([\![\Gamma \vdash N]\!](\vec{d}(i)))$ this amounts to showing

$$\underline{\lambda}i{\in}W. \ [\![\Gamma \vdash M]\!](\vec{d}(i))([\![\Gamma \vdash N]\!](\vec{d}(i))) \in R_\tau$$

which, however, follows immediately from the definition of $R_{\sigma \to \tau}$ because we have $\underline{\lambda}i{\in}W. \ [\![\Gamma \vdash M]\!](\vec{d}(i)) \in R_{\sigma \to \tau}$ and $\underline{\lambda}i{\in}W. \ [\![\Gamma \vdash N]\!](\vec{d}(i)) \in R_\sigma$ by the induction hypotheses on M and N. $\qquad\square$

Thus, for closed λ-terms we get that in particular

Corollary 7.3 *If R is a logical relation of arity W and M is a closed λ–term of type σ then $\underline{\lambda}i{\in}W. \ [\![M]\!] \in R_\sigma$.*

Proof. Immediate from Theorem 7.2 specializing to empty contexts. $\square$

which motivates the following

Definition 7.4 (*R-invariant*)
Let R be a logical relation of arity W. Then an object $d \in D_\sigma$ is called *R-invariant* iff $\delta_W(d) := \underline{\lambda}i{\in}W.d \in R_\sigma$. $\qquad\diamond$

Thus Corollary 7.3 can be reformulated as follows: *the denotation of a closed λ-term is R-invariant for all logical relations R.*

Corollary 7.5 *Let R be a logical relation on the Scott model of arity W and $x_1{:}\sigma_1, \ldots, x_n{:}\sigma_n \vdash M : \tau$ a λ-term. Then $[\![\Gamma \vdash M]\!](\vec{d})$ is R-invariant whenever all d_i in $\vec{d}$ are R-invariant.*

Proof. If all d_i are R–invariant then all $\delta_W(d_i) \in R_{\sigma_i}$ and, therefore,

$$\delta_W(\llbracket \Gamma \vdash M \rrbracket(\vec{d})) = \underline{\lambda} i{\in}W.\, \llbracket \Gamma \vdash M \rrbracket(\langle \delta_W(d_1)(i), \ldots, \delta_W(d_n)(i)\rangle) \in R_\tau$$

by Theorem 7.2. □

Thus, an element of the Scott model is R-invariant if it is λ-definable from elements which are R-invariant. Accordingly, the interpretation of a closed PCF-term is R-invariant if the interpretations of the terms

$$\mathsf{zero} \quad \lambda x{:}\iota.\mathsf{succ}(x) \quad \lambda x{:}\iota.\mathsf{pred}(x) \quad \lambda x{:}\iota.\lambda y{:}\iota.\lambda z{:}\iota.\mathsf{ifz}(x, y, z) \quad \lambda f{:}\sigma{\to}\sigma.\mathsf{Y}_\sigma(f)$$

are all R-invariant in which case we say that "all PCF constants are R-invariant".

We now discuss a property of logical relations guaranteeing that the interpretations of all $\lambda f{:}\sigma{\to}\sigma.\mathsf{Y}_\sigma(f)$ are R-invariant.

Definition 7.6 A logical relation R of arity W is called *admissible* iff $\delta_W(\bot) \in R_\iota$ and R_ι is closed under suprema of directed subsets. ◇

Notice that for finite W there are no nontrivial directed subsets of D_ι^W and, accordingly, in this case a logical relation of arity W is admissible if and only if $\delta_W(\bot) \in R_\iota$.

Theorem 7.7 *Let R be a an admissible logical relation on the Scott model of arity W. Then for all types σ we have that*

(1) *$\delta_W(\bot) \in R_\sigma$ and R_σ is closed under suprema of directed subsets and*
(2) *the interpretation of $\lambda f{:}\sigma{\to}\sigma.\mathsf{Y}_\sigma(f)$ is R-invariant.*

Proof. First we show claim (1) by induction on the structure of σ. For ι claim (1) holds by definition of admissibility. Suppose as induction hypotheses that claim (1) holds for σ and τ. That $\delta_W(\bot) \in R_{\sigma\to\tau}$ follows from $\delta_W(\bot) \in R_\tau$ as insured by the induction hypothesis for τ. Suppose that F is a directed subset of $R_{\sigma\to\tau}$. For showing that $\bigsqcup F \in R_{\sigma\to\tau}$ assume that $d \in R_\sigma$. Then $\underline{\lambda} i{\in}W.f(i)(d(i)) \in R_\tau$ for all $f \in F$ and, therefore, also $\bigsqcup_{f\in F} \underline{\lambda} i{\in}W.f(i)(d(i)) \in R_\tau$ as R_τ is closed under suprema of directed sets by induction hypothesis for τ. Thus, as $\bigsqcup_{f\in F} \underline{\lambda} i{\in}W.f(i)(d(i)) = \underline{\lambda} i{\in}W.\bigsqcup_{f\in F} f(i)(d(i)) = \underline{\lambda} i{\in}W.(\bigsqcup F)(i)(d(i))$ it follows that

$$\underline{\lambda} i{\in}W.\Big(\bigsqcup F\Big)(i)(d(i)) \in R_\tau$$

as desired.

For claim (2) suppose that $f \in R_{\sigma \to \sigma}$. Then one easily shows by induction that $\underline{\lambda}i{\in}W.f(i)^n(\bot) \in R_\sigma$ for all $n \in \mathbb{N}$. As by (1) the set R_σ is closed under suprema of directed sets it follows that

$$\underline{\lambda}i{\in}W.\delta_W([\![\lambda f{:}\sigma{\to}\sigma.\mathsf{Y}_\sigma(f)]\!])(i)(f(i)) = \underline{\lambda}i{\in}W.\mu(f(i)) \in R_\sigma$$

as desired. $\square$

This has the following immediate consequence.

Theorem 7.8   *Let R be an admissible logical relation on the Scott model such that the interpretations of the terms*

zero $\lambda x{:}\iota.\mathsf{succ}(x)$ $\lambda x{:}\iota.\mathsf{pred}(x)$ $\lambda x{:}\iota.\lambda y{:}\iota.\lambda z{:}\iota.\mathsf{ifz}(x, y, z)$

are all R-invariant. Then all interpretations of closed PCF-terms are R-invariant.

Proof.   Immediate from (the remark after) Corollary 7.5 and Theorem 7.7(2). $\square$

We now consider some (useful) examples of logical relations satisfying the premises of Theorem 7.8.

$$(x, y, z) \in R_\iota^{(1)} \quad \Leftrightarrow \quad x{\uparrow}y \wedge z = x \sqcap y$$
$$(x, y, z) \in R_\iota^{(2)} \quad \Leftrightarrow \quad x{=}\bot \vee y{=}\bot \vee z{=}\bot \vee x{=}y{=}z$$

where $x{\uparrow}y$ is an abbreviation for $\exists z.\, x \sqsubseteq z \wedge y \sqsubseteq z$.

As the arity of these relations is finite and $(\bot, \bot, \bot)$ is an element of both $R_\iota^{(1)}$ and $R_\iota^{(2)}$ it follows that they are admissible. That the interpretations of zero, $\lambda x{:}\iota.\mathsf{succ}(x)$, $\lambda x{:}\iota.\mathsf{pred}(x)$ and $\lambda x{:}\iota.\lambda y{:}\iota.\lambda z{:}\iota.\mathsf{ifz}(x, y, z)$ are all invariant under both $R^{(1)}$ and $R^{(2)}$ is a straightforward, but tedious exercise.

Thus, we can now give the

Proof (of Lemma 6.3) :
As the logical relation $R^{(1)}$ satisfies the premises of Theorem 7.8 we know that the interpretation of every closed PCF term of first order type is $R^{(1)}$-invariant and thus stable. $\square$

As $R^{(2)}$ satisfies the premises of Theorem 7.8 we know that every PCF definable $f \in D_{\iota \to \iota \to \iota}$ is $R^{(2)}$-invariant. Such an f cannot simultaneously satisfy $f0\bot = 0 = f\bot 0$ and $f11 = 1$ as $(\bot, 0, 1)$ and $(0, \bot, 1)$ are in $R_\iota^{(2)}$ whereas $(f\bot 0, f0\bot, f11) = (0, 0, 1)$ is not. This consideration provides an alternative proof of Lemma 6.3.

However, we have that

Lemma 7.9 *Plotkin's continuous existential quantifier is $R^{(2)}$-invariant.*

Proof. Suppose $(\exists(f_1), \exists(f_2), \exists(f_3)) \notin R_\iota^{(2)}$ then w.l.o.g. one of the following to cases applies

(1) $\exists(f_1) = 0 = \exists(f_2)$ and $\exists(f_3) = 1$
(2) $\exists(f_1) = 0$ and $\exists(f_2) = 1 = \exists(f_3)$.

In case (1) there exist $n_1, n_2 \in \mathbb{N}$ with $f_1(n_1) = 0 = f_2(n_2)$ and $f_3(\bot) = 1$. As $(n_1, n_2, \bot) \in R_\iota^{(2)}$ it follows that $(f_1, f_2, f_3) \notin R_{\iota \to \iota}^{(2)}$.
In case (2) there exists an $n \in \mathbb{N}$ with $f_1(n) = 0$ and $f_2(\bot) = 1 = f_3(\bot)$. As $(n, \bot, \bot) \in R_\iota^{(2)}$ it follows that $(f_1, f_2, f_3) \notin R_{\iota \to \iota}^{(2)}$. $\qquad \square$

As $\exists$ is $R^{(2)}$-invariant but por is not $R^{(2)}$-invariant it follows from Corollary 7.5 that por is not PCF definable from $\exists$.

More generally, there arises the question to which extent one can characterise PCF definability via logical relations. To some extent K. Sieber has achieved such a characterization up to type level ≤ 2, i.e. for types $\sigma_1 \to \ldots \to \sigma_n \to \iota$ where the σ_i are all first order, in [Sieber 1992] where he has shown that for types σ of type level ≤ 2 an element $d \in D_\sigma$ arises as supremum of a directed set of PCF-definable elements if and only if d is invariant under all logical relations of finite arity satisfying the premisses of Theorem 7.8. Moreover, he has given also a purely combinatorial, syntax-free characterization of this class of logical relations. However, using a wider class of logical relations (of varying arity also called *Kripke logical relations*) one may characterise for arbitrary types σ those $d \in D_\sigma$ which arise as suprema of directed sets of PCF-definable elements. Later in Chapters 10 and 11 we will use Kripke logical relations for constructing a fully abstract model of PCF in a completely syntax-free way.

In his seminal paper [Plotkin 1977] G. Plotkin has shown that $\exists$ is not definable from por by a purely syntactical argument. The following consideration shows that this cannot be achieved by an argument using admissible logical relations. Consider

$$\exists_n(f : \iota \to \iota) = \mathsf{ifz}(f(\underline{0})\, \mathsf{por} \ldots \mathsf{por} f(\underline{n}), \underline{0}, f(\Omega_\iota))$$

which is obviously definable in PCF+por. As $\exists$ is the supremum of the increasing chain $(\exists_n)_{n \in \mathbb{N}}$ it follows that $\exists$ is invariant under all admissible logical relations under which the constants of PCF and por are invariant.

Chapter 8

Some Structural Properties of the D_σ

Up to now we know about the D_σ arising in the Scott model of PCF just that they are domains, i.e. have suprema of directed subsets and a least element. Actually, they have much more properties which will be investigated in this chapter.

Lemma 8.1 *For every* PCF *type σ the cpo D_σ has infima of nonempty finite subsets and there exists a closed* PCF *term glb^σ of type $\sigma\to\sigma\to\sigma$ such that $[\![\mathsf{glb}^\sigma]\!]xy$ is the infimum of x and y for all $x, y \in D_\sigma$.*

Proof. For ι we may define glb^ι as $\lambda x{:}\iota.\lambda y{:}\iota.\mathsf{ifz}(\mathsf{eq}xy, x, \Omega_\iota)$ where eq is a PCF term deciding equality of natural numbers. The function eq can for example be implemented by the PCF term $\lambda x{:}\iota.\lambda y{:}\iota.(x \dot- y)+(y \dot- x)$ where $\dot-$ stands for truncated subtraction. Obviously, $[\![\mathsf{glb}^\iota]\!]xy = x$ if $x = y$ and $\bot$ otherwise and, therefore, delivers the binary infimum of x and y as desired.

Now if by induction hypothesis there is a PCF term glb^σ computing the binary infimum for D_σ then

$$\mathsf{glb}^{\tau\to\sigma} \equiv \lambda f{:}\tau\to\sigma.\lambda g{:}\tau\to\sigma.\lambda y{:}\tau.\,\mathsf{glb}^\sigma(f(y), g(y))$$

computes the binary infimum in $D_{\tau\to\sigma}$ which can be seen as follows. If f and g are continuous functions from D_τ to D_σ then $[\![\mathsf{glb}^{\tau\to\sigma}]\!]fg$ is continuous and $[\![\mathsf{glb}^{\tau\to\sigma}]\!]fgy = f(y)\sqcap g(y)$ for all $y \in D_\tau$. Obviously, we have $[\![\mathsf{glb}^{\tau\to\sigma}]\!]fg \sqsubseteq f, g$ and if $h : D_\tau\to D_\sigma$ with $h \sqsubseteq f, g$ then $h(y) \sqsubseteq f(y) \sqcap g(y)$ for all $y \in D_\tau$, i.e. $h \sqsubseteq [\![\mathsf{glb}^{\tau\to\sigma}]\!]fg$. Thus $[\![\mathsf{glb}^{\tau\to\sigma}]\!]fg$ is the infimum of f and g as desired. $\qquad\square$

Next we will show that every element in D_σ appears as supremum of an ascending chain of "finite approximations".

Definition 8.2   Let less be a PCF term of type $\iota\to\iota\to\iota$ with

$$[\![\text{less}]\!]xy = \begin{cases} 0 & \text{if } x,y \in \mathbb{N} \text{ and } x < y \\ 1 & \text{if } x,y \in \mathbb{N} \text{ and } x \geq y \\ \bot & \text{otherwise.} \end{cases}$$

For all PCF types σ we define closed PCF terms ψ_n^σ of type $\sigma\to\sigma$ by recursion on the structure of σ via the following two clauses

$$\psi_n^\iota \equiv \lambda x{:}\iota.\,\text{ifz}(\text{less } x\,\underline{n}, x, \Omega_\iota)$$
$$\psi_n^{\sigma\to\tau} \equiv \lambda f{:}\sigma\to\tau.\lambda x{:}\sigma.\psi_n^\tau(f(\psi_n^\sigma(x)))$$

for all $n\in\mathbb{N}$.   ◇

The next lemma identifies characteristic properties of the functions denoted by the ψ_n^σ.

Lemma 8.3   *For $h_n^\sigma := [\![\psi_n^\sigma]\!]$ it holds that*

(1) $h_n^\sigma \circ h_n^\sigma = h_n^\sigma \sqsubseteq \text{id}_{D_\sigma}$
(2) $(h_n^\sigma)_{n\in\mathbb{N}}$ *is an ascending chain whose supremum is* id_{D_σ}
(3) $h_n^\sigma[D_\sigma] := \{h_n^\sigma(d)\,|\,d \in D_\sigma\}$ *is finite.*

Proof.   Obviously, the three requirements hold for ι.

Suppose as induction hypotheses that the three requirements hold for σ and τ. Obviously, we have $h_n^{\sigma\to\tau}(f) = h_n^\tau \circ f \circ h_n^\sigma$ from which (1) and (2) for $\sigma\to\tau$ follow easily from the requirements (1) and (2) for σ and τ as ensured by the induction hypotheses. Furthermore, from $h_n^{\sigma\to\tau}(f) = h_n^\tau \circ f \circ h_n^\sigma$ it follows[1] that $\left|h_n^{\sigma\to\tau}[D_{\sigma\to\tau})]\right| \leq \left|h_n^\tau[D_\tau]\right|^{\left|h_n^\sigma[D_\sigma]\right|}$ which is finite as by induction hypothesis $\left|h_n^\sigma[D_\sigma]\right|$ and $\left|h_n^\tau[D_\tau]\right|$ are finite.   □

Now we will study the properties identified in Lemma 8.1 and Lemma 8.3 axiomatically as requirements for cpo's.

Definition 8.4   A cpo D is called SFP iff there exists an ascending chain $(h_n : D\to D)_{n\in\mathbb{N}}$ of continuous functions such that

(1) every h_n is a projection, i.e. $h_n \circ h_n = h_n \sqsubseteq \text{id}_D$
(2) $\bigsqcup_{n\in\mathbb{N}} h_n = \text{id}_D$
(3) every h_n is finitary, i.e. $h_n[D]$ is finite.

Accordingly, a cpo D is called SFP iff id_D is the supremum of an ascending chain of finitary projections.   ◇

---

[1] as $h_n^{\sigma\to\tau}(f) = h_n^{\sigma\to\tau}(g)$ are equal iff their restrictions to $h_n^\sigma[D_\sigma]$ are equal

If id_D is the supremum of a chain h_n of finitary projections then we may call an element $d \in D$ *compact* (or simply *finite*) iff $d = h_n(d)$ for some $n \in \mathbb{N}$. The next lemma shows that this notion is independent from the choice of the sequence $(h_n)_{n\in\mathbb{N}}$ exhibiting D as an SFP cpo.

Lemma 8.5   *Let $(h_n{:}D{\rightarrow}D)_{n\in\mathbb{N}}$ be an increasing sequence of finitary projections whose supremum is id_D. Then for $e \in D$ the following two conditions are equivalent*

(1) $e = h_n(e)$ *for some $n \in \mathbb{N}$*
(2) *for every directed subset $X \subseteq D$ with $e \sqsubseteq \bigsqcup X$ there exists an $x \in X$ with $e \sqsubseteq x$.*

Proof.   Suppose $e = h_n(e)$. Let X be a directed subset of D with $e \sqsubseteq \bigsqcup X$. Then $e = h_n(e) \sqsubseteq h_n(\bigsqcup X) = \bigsqcup h_n[X]$. As $h_n[X]$ is finite (since $h_n[D]$ is finite) and directed there exists an $x \in X$ with $h_n(x) = \bigsqcup h_n[X] \sqsupseteq e$. For such an x we have $e \sqsubseteq h_n(x) \sqsubseteq x \in X$ as desired.

For the reverse direction suppose that e satisfies (2). Notice that the set $\{h_n(e) \mid n \in \mathbb{N}\}$ is directed and its supremum is e. Thus, as e satisfies (2) there exists an $n \in \mathbb{N}$ with $e \sqsubseteq h_n(e)$. As $h_n(e) \sqsubseteq e$ it follows that $e = h_n(e)$ as desired.   □

Notice that condition (2) of the previous lemma makes sense for arbitrary cpo's (and not only for those satisfying the SFP property).

Definition 8.6   (compact elements)
Let A be a cpo. An element $e \in A$ is called *compact* (or *finite*) iff for every directed subset X of A with $e \sqsubseteq \bigsqcup X$ there exists already an $x \in X$ with $e \sqsubseteq x$. We write $\mathcal{K}(A)$ for the set of compact elements of A.   ◇

Lemma 8.7   *Let A be a cpo with the property SFP. Then for every $a \in A$ the set $\mathcal{K}_a := \{e \in \mathcal{K}(A) \mid e \sqsubseteq a\}$ is directed and $a = \bigsqcup \mathcal{K}_a$.*

Proof.   Let (h_n) be a chain of finitary projections whose supremum is id_A and $a \in A$. First we show that $\mathcal{K}_a$ is directed. For that purpose suppose that $e_1, e_2 \in \mathcal{K}_a$. Thus, since $\{h_n(a) \mid n \in \mathbb{N}\}$ is directed and has supremum a there exists $n_1, n_2 \in \mathbb{N}$ with $e_i \sqsubseteq h_{n_i}(a)$ for $i{=}1,2$. Then for $n := \mathsf{max}(n_1, n_2)$ we have $e_1, e_2 \sqsubseteq h_n(a)$. As by Lemma 8.5 we have $h_n(a) \in \mathcal{K}(A)$ it is an element of $\mathcal{K}_a$ above e_1 and e_2. Thus $\mathcal{K}_a$ is directed. As $\{h_n(a) \mid n \in \mathbb{N}\}$ is a subset of $\mathcal{K}_a$ and $a = \bigsqcup h_n(a)$ it follows that $a = \bigsqcup \mathcal{K}_a$ as desired.   □

Next we will show that elements of an SFP predomain A are in 1-1-correspondence with order-theoretic ideals in the poset $\mathcal{K}(A)$.

Theorem 8.8   *Let A be an* SFP *predomain. A subset I of $\mathcal{K}(A)$ is called an* ideal *iff I is downward closed and directed. We write $\mathsf{Idl}(\mathcal{K}(A))$ for the poset of ideals in $\mathcal{K}(A)$ ordered by $\subseteq$. Then the map*

$$i_A : A \to \mathsf{Idl}(\mathcal{K}(A)) : a \mapsto \mathcal{K}_a$$

is an isomorphism of posets.

Proof.   Obviously, the map i_A is monotonic. It reflects the order as if $\mathcal{K}_{a_1} \subseteq \mathcal{K}_{a_2}$ then $a_1 = \bigsqcup \mathcal{K}_{a_1} \sqsubseteq \bigsqcup \mathcal{K}_{a_2} = a_2$ by Lemma 8.7. As i_A reflects the order it is one-to-one. Thus, it remains to show that i_A is surjective. Suppose that I is an ideal in $\mathcal{K}(A)$. Let $a := \bigsqcup I$. We show that $I = \mathcal{K}_a$. Obviously, we have $I \subseteq \mathcal{K}_a$. But if $e \in \mathcal{K}_a$ then (as e is compact) there is an $e' \in I$ with $e \sqsubseteq e'$ from which it follows that $e \in I$ since I is downward closed. $\square$

Next we show that SFP predomains are closed under a lot of useful constructions.

Theorem 8.9   SFP *predomains are closed under $\times$ and $\to$ and for a set S the domain $S_\perp$ is* SFP *iff S is countable.*

Proof.   Let A and B be SFP predomains. Then there exist ascending chains $(h_n)_{n \in \mathbb{N}}$ and $(k_n)_{n \in \mathbb{N}}$, respectively, of finitary projections with $\mathsf{id}_A = \bigsqcup_{n \in \mathbb{N}} h_n$ and $\mathsf{id}_B = \bigsqcup_{n \in \mathbb{N}} k_n$.

The predomain $A \times B$ is SFP as

$$h_n \times k_n : A \times B \to A \times B : (a, b) \mapsto (h_n(a), k_n(b))$$

is a finitary projection for all $n \in \mathbb{N}$ and $\mathsf{id}_{A \times B} = \mathsf{id}_A \times \mathsf{id}_B = \bigsqcup_{n \in \mathbb{N}} h_n \times k_n$.

The predomain $[A \to B]$ is SFP as

$$h_n \to k_n : [A \to B] \to [A \to B] : f \mapsto k_n \circ f \circ h_n$$

is a finitary projection for all $n \in \mathbb{N}$ and $\mathsf{id}_{A \to B} = \mathsf{id}_A \to \mathsf{id}_B = \bigsqcup_{n \in \mathbb{N}} h_n \to k_n$. That the image of $h_n \to k_n$ is finite follows from the facts that there are just finitely many functions from $h_n[A]$ to $k_n[B]$ and that $k_n \circ f \circ h_n$ is determined uniquely by its restriction to $h_n[A]$.

For a set S all elements of $S_\perp$ are compact. If $S_\perp$ is SFP then $S_\perp = \mathcal{K}(S_\perp)$ is countable and, therefore, the set S itself has to be countable. On the other hand if S is countable then one may enumerate its elements as $s_0, s_1, \dots, s_n, \dots$. Define h_n as the mapping sending s_i with $i < n$ to s_i and all other arguments to $\perp$. Obviously, the h_n form an ascending sequence of finitary projections whose supremum is $\mathsf{id}_{S_\perp}$. $\square$

Thus the D_σ and their finite cartesian products are all SFP domains and accordingly determined by their subposets of compact elements as ensured by Theorem 8.8.

It is a straightforward exercise(!) to show that SFP predomains are also closed under $(-)_\perp$ (lifting[2]) and $+$ and that SFP domains are closed under $\oplus$ (coalesced sum[3]), $\otimes$ (smash product[4].) and $\circ\!\!\rightarrow$ (strict function space[5]).

Next we show that continuous functions from SFP predomains A to arbitray cpo's B are (by restriction to $\mathcal{K}(A)$) in 1-1-correspondence with the monotonic maps from $\mathcal{K}(A)$ to B.

Theorem 8.10 *Let A be an SFP predomain and B an arbitrary cpo. Then every continuous map $f : A \to B$ is uniquely determined by its restriction to $\mathcal{K}(A)$ and every monotonic map $h : \mathcal{K}(A) \to B$ extends to a continuous map $f : A \to B$, i.e. there is a unique continuous $f : A \to B$ with $h = f{\restriction}\mathcal{K}(A)$.*

Proof. Suppose f and g are continuous maps from A to B with $f{\restriction}\mathcal{K}(A) = g{\restriction}\mathcal{K}(A)$. We have to show that for an arbitrary $a \in A$ it holds that $f(a) = g(a)$. For all $e \in \mathcal{K}_a$ we have $f(e) = g(e)$ since $f{\restriction}\mathcal{K}(A) = g{\restriction}\mathcal{K}(A)$ and, therefore, we have $f(a) = \bigsqcup_{e\in\mathcal{K}_a} f(e) = \bigsqcup_{e\in\mathcal{K}_a} g(e) = g(a)$ as desired.

Suppose $h : \mathcal{K}(A) \to B$ is monotonic. Its tentative continuous extension f is defined by putting $f(a) = \bigsqcup h[\mathcal{K}_a]$. Obviously, the map f is monotonic and $f(e) = h(e)$ for compact e in A. For showing that f is also continuous consider a directed set $X \subseteq A$. As f is monotonic it suffices to show that $f(\bigsqcup X) \sqsubseteq \bigsqcup f(X)$. From the definition of f we know that $f(\bigsqcup X)$ is the supremum of all $h(e)$ with e compact and $e \sqsubseteq \bigsqcup X$. But a compact e is below $\bigsqcup X$ if and only if $e \sqsubseteq x$ for some $x \in X$. Thus we get that $f(\bigsqcup X)$ is the supremum of all $h(e)$ where e is compact and $e \sqsubseteq x$ for some $x \in X$. As this latter supremum is below $\bigsqcup f(X)$ it follows that $f(\bigsqcup X) \sqsubseteq \bigsqcup f(X)$ as desired. $\square$

Finally, we introduce the notion of a *Scott domain* and show that, in particular, all D_σ and their finite products are actually Scott domains.

---

[2]The "lifting" $A_\perp$ of a predomain A is the domain obtained from A by adding a new bottom element.

[3]$A \oplus B$ is obtained from the disjoint union $A + B$ by identifying its two minimal elements.

[4]$A \otimes B$ is obtained from $A \times B$ by identifying all pairs where at least one component equals $\perp$.

[5]$[A\circ\!\!\rightarrow B]$ is obtained from $[A\to B]$ by removing all non-strict maps.

Definition 8.11   A cpo A has *continuous binary infima* iff for all $x, y \in A$ their infimum $x \sqcap y$ exists and the function

$$\sqcap : A \times A \to A : (x, y) \mapsto x \sqcap y$$

is Scott continuous. A *Scott (pre)domain* is an SFP (pre)domain with continuous binary infima. $\Diamond$

Obviously, a cpo A has continuous binary infima iff it has binary infima and $\sqcap$ satisfies the following restricted distributivity law, namely

$$a \sqcap \bigsqcup_{x \in X} x = \bigsqcup_{x \in X} a \sqcap x$$

for all $a \in A$ and all *directed* $X \subseteq A$.

For showing that Scott domains are closed under the usual type forming operations we need besides Theorem 8.9 the following lemma.

Lemma 8.12   *Cpo's with continuous binary infima are closed under $\times$ and $\to$. Moreover, for all sets S the cpo $S_\perp$ has binary continuous infima.*

Proof.   That $S_\perp$ has continuous binary infima is an easy exercise.

Suppose that A and B are cpo's having continuous binary infima. That $A \times B$ has continuous binary infima follows from the fact that

$$(x_1, y_1) \sqcap (x_2, y_2) = (x_1 \sqcap_A x_2, y_1 \sqcap_B y_2)$$

and $\sqcap_A$ and $\sqcap_B$ are continuous by assumption. That $[A \to B]$ has continuous binary infima can be seen as follows. Given $f, g \in [A \to B]$ the function $f \sqcap g = \underline{\lambda} x{:}A.f(x) \sqcap_B g(x)$ is continuous as $\sqcap_B$ is continuous and one easily sees that $f \sqcap g$ is the infimum of f and g. That the binary function $\sqcap$ on $[A \to B]$ is continuous follows from the fact that it is λ-definable from the continuous function $\sqcap_B$. $\square$

Theorem 8.13   *Scott domains are closed under $\times$ and $\to$ and $S_\perp$ is a Scott domain for all countable sets S.*

Proof.   Immediate from Theorem 8.9 and Lemma 8.12. $\square$

Moreover, it is an easy exercise(!) to show that Scott domains are closed under the further domain constructions $(-)_\perp$, separated sum $\big((-_1) + (-_2)\big)_\perp$, $\oplus$, $\otimes$ and $\circ\!\!\to$.

Notice that our definition of Scott domain is somewhat unorthodox but equivalent to the usual one (as e.g. in [Griffor et.al. 1994]) as shown in the next theorem.

Theorem 8.14 *A cpo D with $\bot$ is a Scott domain iff D is*

(1) bounded complete, *i.e. all bounded subsets of D have a supremum in D, and*
(2) countably algebraic, *i.e. $\mathcal{K}(D)$ is countable and for every $d \in D$ the set $\mathcal{K}_d = \{e \in \mathcal{K}(D) \mid e \sqsubseteq d\}$ is directed and has supremum d.*

Proof. Suppose D is a Scott domain. Since D is SFP it is countably algebraic. For showing that D is bounded complete it suffices to show that every finite bounded set of compact elements has a supremum in D.

Suppose $e_1, \ldots, e_n$ are compact elements of D having a common upper bound b in D. Since D is SFP there exists a finitary projection $h : D \to D$ with $e_1, \ldots, e_n \in h[D]$. Obviously, the elements $e_1, \ldots, e_n$ are bounded in $h[D]$ by $h(b)$. Thus $\{e_1, \ldots, e_n\}$ has a minimal upper bound in $h[D]$ because $h[D]$ is finite. Suppose e' and e'' are minimal upper bounds of $\{e_1, \ldots, e_n\}$ in $h[D]$. Then for $i \in \{1, \ldots, n\}$ we have $e_i = h(e_i) \sqsubseteq h(e' \sqcap e'') \sqsubseteq e', e''$ from which it follows that $e' = e''$. Now let e be the supremum of $e_1, \ldots, e_n$ in $h[D]$. Suppose d is an upper bound of $e_1, \ldots, e_n$ in D. Then $h(d)$ is an upper bound of $e_1, \ldots, e_n$ in $h[D]$ and thus $e \sqsubseteq h(d) \sqsubseteq d$. Thus e is the supremum of $e_1, \ldots, e_n$ in D.

Suppose D is bounded complete and countably algebraic. Let $\{e_n \mid n \in \mathbb{N}\}$ be an enumeration of $\mathcal{K}(D)$. First observe that suprema of bounded finite subsets of $\mathcal{K}(D)$ are compact (exercise!). Let D_n be the least subset of D which is closed under finite suprema and contains all e_i with $i < n$. Obviously D_n is a finite set of compact elements. Let $h_n : D \to D : d \mapsto \bigsqcup\{e \in D_n \mid e \sqsubseteq d\}$. One easily shows (exercise!) that all h_n are finitary projections and $\mathrm{id}_D = \bigsqcup_{n \in \mathbb{N}} h_n$ thus exhibiting D as an SFP domain.

It remains to show that D has binary continuous infima. Suppose $x, y \in D$. Then the set $\{z \in D \mid z \sqsubseteq x, y\}$ is bounded and thus has a supremum giving rise to $x \sqcap y$. For showing continuity of $\sqcap : D \times D \to D$ suppose that $X \subseteq D$ is directed and $y \in D$. Obviously, we have $y \sqcap x \sqsubseteq y \sqcap \bigsqcup X$ for all $x \in X$ and thus $\bigsqcup_{x \in X} y \sqcap x \sqsubseteq y \sqcap \bigsqcup X$. For the reverse direction suppose $e \in \mathcal{K}(D)$ with $e \sqsubseteq y \sqcap \bigsqcup X$. Then $e \sqsubseteq y$ and $e \sqsubseteq \bigsqcup X$. Since e is compact there exists $x \in X$ with $e \sqsubseteq x$. Thus, we have $e \sqsubseteq y \sqcap x \sqsubseteq \bigsqcup_{x \in X} y \sqcap x$. Since this implication holds for all $e \in \mathcal{K}(D)$ and D is countably algebraic we conclude that $y \sqcap \bigsqcup X \sqsubseteq \bigsqcup_{x \in X} y \sqcap x$ as desired. $\square$

Notice that the above proof shows in particular that in algebraic domains binary infima are continuous provided they always exist.

The motivation for our definition of Scott domain is that it appears as the most natural one when starting to investigate the structure of the domains arising in the Scott model of PCF. The usual definition, however, is motivated by weakening the most respectable notion of a "countably algebraic lattice", i.e. a complete lattice satisfying condition (2). As one wants to get rid of the annoying $\top$ element which does not have a computational meaning it appears as most natural to weaken completeness by requiring (besides directed completeness) the existence of suprema just for bounded sets and not for arbitrary subsets.

Notice, however, that the class of Scott domains is a bit wider than actually needed because all domains showing up in Scott semantics of programming languages actually satisfy the following stronger requirement.

Definition 8.15   A cpo A is called *coherently complete* iff all coherent subsets X of A have a supremum in A where a subset X of A is called *coherent* iff all $x, y \in X$ have an upper bound in A. $\diamondsuit$

The reader is invited to show that all D_σ are coherently complete. Actually, one can show that coherently complete domains are closed under all the usual domain constructions.[6]

An aesthetically pleasing aspect of coherently complete domains is that they can be characterised as those partial orders where every coherent subset has a supremum (because every directed set is coherent).

---

[6]One exception is the so-called *Smyth powerdomain*. However, Scott domains are *not* closed under the *Plotkin powerdomain* construction whereas SFP domains are closed under this latter construction and were introduced by G. Plotkin (see [Plotkin 1978]) precisely for this purpose. Thus, if one wants to have closure under (all sorts of) powerdomains then one should work with SFP domains and otherwise coherently complete SFP domains are absolutely sufficient.

Chapter 9

Solutions of Recursive Domain Equations

Unlike PCF "real" functional programming languages like ML or Haskell provide the facility of defining types recursively. A recursive definition of type A takes the form of a "domain equation" $A = E[A]$ where the right hand side is a type expression typically involving the recursively defined type A. Typical examples of such domain equations are

$$N = 1_\perp \oplus N \qquad S = A \otimes S_\perp \qquad D = N \oplus [D {\to} D]_\perp \qquad C = R^C {\times} C$$

where $(-)_\perp$ stands for lifting, $\oplus$ for coalesced sum and $\otimes$ for the so-called "smash product"[1]. Although we use the symbol $=$ in domain equations we rather mean $\cong$, i.e. that the domain on the left hand side should be isomorphic to the domain on the right hand side of the domain equation. The intended solution for the domain equation for N is $\mathbb{N}_\perp$. If A is a flat domain $M_\perp$ (where M is a set of "tokens") then the intended solution for the domain equation for S is the domain of "streams over M, i.e. the finite and infinite sequences of elements of M under the prefix-ordering. The intended solution of the domain equation for D is not so easy to describe but the intention is that the elements of D different from $\perp_D$ are either natural numbers or continuous functions from D to D (where $\perp_D$ is distinguished from the function $\underline{\lambda}d{:}D.\perp_D$). Solutions of $C = R^C {\times} C$ are also not so easy to visualize or describe in a concrete way but notice that for any such C we have that $R^C \cong R^{R^C \times C} \cong (R^C)^{(R^C)}$, i.e. we get a nontrivial solution of the domain equation $D = D^D$ when R is nontrivial.

The general form of a domain equation is $D = F(D, D)$ where F is a *locally continuous* functor from $\mathcal{C}^{op} {\times} \mathcal{C}$ to $\mathcal{C}$ and $\mathcal{C}$ is the category of domains and *strict* functions. Here "locally continuous" means that the function

---

[1]If A and B are domains then their smash product $A \otimes B$ consists of all pairs $(a, b) \in A {\times} B$ with $a{=}\perp \vee b{=}\perp \Rightarrow a{=}\perp{=}b$ and the order on $A{\otimes}B$ is inherited from $A{\times}B$.

$F : \mathcal{C}(Y_2, Y_1) \times \mathcal{C}(X_1, X_2) \to \mathcal{C}(F(Y_1, X_1), F(Y_2, X_2))$ is Scott continuous for all objects X_1, X_2, Y_1, Y_2 in $\mathcal{C}$. This assumption is guaranteed when the right hand side of a domain equation is built up from 1 (containing just $\perp$) by the functors $(-)_\perp$ (lifting), $\times$ (cartesian product), $\otimes$ (smash product), $+$ (separated sum), $\oplus$ (coalesced sum), $\to$ (function space) and $\circ\!\to$ (strict function space) where only the last two make proper use of their contravariant argument.

Definition 9.1   Let $F : \mathcal{C}^{op} \times \mathcal{C} \to \mathcal{C}$ be a locally continuous functor. A *bifree solution of* $X = F(X, X)$ is a domain D together with an isomorphism $\alpha : F(D, D) \to D$ such that every strict $e : D \to D$ with $e = \alpha \circ F(e, e) \circ \alpha^{-1}$ is equal to id_D. $\Diamond$

We show now that bifree solutions are unique up to isomorphism.

Lemma 9.2   Let $F : \mathcal{C}^{op} \times \mathcal{C} \to \mathcal{C}$ be a *locally continuous functor. If* $\alpha : F(A, A) \to A$ and $\beta : F(B, B) \to B$ are bifree solutions of the domain equation $X = F(X, X)$ then there exists a unique isomorphism $i : A \to B$ with $i = \beta \circ F(i^{-1}, i) \circ \alpha^{-1}$.

Proof.   Let $(i : A \to B, j : B \to A)$ be the least solution of the equations

$$i = \beta \circ F(j, i) \circ \alpha^{-1} \qquad\qquad j = \alpha \circ F(i, j) \circ \beta^{-1}$$

which exists as the assignment $(i, j) \mapsto (\beta \circ F(j, i) \circ \alpha^{-1}, \alpha \circ F(i, j) \circ \beta^{-1})$ gives rise to a continuous function $\varphi = \langle \varphi_1, \varphi_2 \rangle : \mathcal{C}(A, B) \times \mathcal{C}(B, A) \to \mathcal{C}(A, B) \times \mathcal{C}(B, A)$. But then we have

$$j \circ i = \alpha \circ F(j \circ i, j \circ i) \circ \alpha^{-1} \qquad\qquad i \circ j = \beta \circ F(i \circ j, i \circ j) \circ \beta^{-1}$$

from which it follows that $j \circ i = \mathrm{id}_A$ and $i \circ j = \mathrm{id}_B$ as both α and β are bifree solutions by assumption. Thus, the map j equals i^{-1} from which it follows that $i = \beta \circ F(i^{-1}, i) \circ \alpha^{-1}$ as desired.

Let $\iota : A \to B$ be some isomorphism with $\iota = \beta \circ F(\iota^{-1}, \iota) \circ \alpha^{-1}$. We want to show that $\iota = i$ and $\iota^{-1} = j$ where i and j are defined as above. For this purpose consider the continuous function $\delta : \mathcal{C}(A, A) \to \mathcal{C}(A, A) :$ $e \mapsto \alpha \circ F(e, e) \circ \alpha^{-1}$. We write e_n for $\delta^n(\perp)$. As α is bifree we have $\mathrm{id}_A = \bigsqcup_{n \in \mathbb{N}} e_n$. We will show now by induction on n that

$$(\iota \circ e_n, e_n \circ \iota^{-1}) = \varphi^n(\perp, \perp)$$

from which it follows that $(\iota, \iota^{-1}) = (i, j)$. For $n{=}0$ we have $(\iota \circ e_0, e_0 \circ \iota^{-1}) = (\perp, \perp)$ since $\iota \circ \perp = \perp$ and $\perp \circ \iota^{-1} = \perp$. Now suppose as induction hypothesis that $(\iota \circ e_n, e_n \circ \iota^{-1}) = \varphi^n(\perp, \perp)$. Then

$$\iota \circ e_{n+1} = \beta \circ F(\iota^{-1}, \iota) \circ \alpha^{-1} \circ \alpha \circ F(e_n, e_n) \circ \alpha^{-1} =$$
$$= \beta \circ F(e_n \circ \iota^{-1}, \iota \circ e_n) \circ \alpha^{-1} = \varphi_1(\varphi^n(\bot, \bot))$$

where the last equality follows from the induction hypothesis and

$$e_{n+1} \circ \iota^{-1} = \alpha \circ F(e_n, e_n) \circ \alpha^{-1} \circ \alpha \circ F(\iota, \iota^{-1}) \circ \beta^{-1} =$$
$$= \alpha \circ F(\iota \circ e_n, e_n \circ \iota^{-1}) \circ \beta^{-1} = \varphi_2(\varphi^n(\bot, \bot))$$

where the last equality follows from the induction hypothesis. Thus, we have

$$(\iota \circ e_{n+1}, e_{n+1} \circ \iota^{-1}) = \langle \varphi_1(\varphi^n(\bot, \bot)), \varphi_2(\varphi^n(\bot, \bot)) \rangle = \varphi^{n+1}(\bot, \bot)$$

as desired. □

Now we will show that for every locally continuous functor $F : \mathcal{C}^{op} \times \mathcal{C} \to \mathcal{C}$ there exists a bifree solution of the domain equation $X = F(X, X)$ which we know to be unique up to isomorphism by the previous Lemma 9.2. However, for this purpose we need some preparatory notions and lemmas.

Definition 9.3 An *embedding/projection pair* from A to B is a pair (e, p) where $e : A \to B$ and $p : B \to A$ are continuous functions with $p \circ e = \mathsf{id}_A$ and $e \circ p \sqsubseteq \mathsf{id}_B$. We call e *embedding* and p *projection*. ◇

One easily sees that for an embedding/projection pair (e, p) from A to B we have $e(a) \sqsubseteq b \Leftrightarrow a \sqsubseteq p(b)$ for all $a \in A$ and $b \in B$ from which it follows that $p(b)$ is the greatest a with $e(a) \sqsubseteq b$. We leave it as an exercise(!) to show that embeddings and projections are always strict.

Next we show that one component of an embedding/projection pair determines the other one uniquely.

Lemma 9.4 Let (e, p) and (e', p') be embedding projection pairs from A to B. Then $(e, p) = (e', p')$ whenever $e = e'$ or $p = p'$.

Proof. If $e = e'$ then we have $p = p' \circ e \circ p \sqsubseteq p'$ and similarly $p' \sqsubseteq p$ from which it follows that $p = p'$. If $p = p'$ then $e = e \circ p \circ e' \sqsubseteq e'$ and similarly $e' \sqsubseteq e$ from which it follows that $e = e'$. □

Accordingly, we say that a map e is an *embedding* iff there is a map p such that (e, p) is an embedding/projection pair and that a map p is a *projection* iff there is a map e such that (e, p) is an embedding/projection pair.

Canonical solutions of domain equations will be constructed as inverse limits of projections. The notion of inverse limit will be explained in the next theorem.

Theorem 9.5   Let $(f_n : D_{n+1} \to D_n \mid n \in \mathbb{N})$ be a sequence of maps in $\mathcal{C}$. Its inverse limit is given by the sequence $q_n : D \to D_n$ where

$$|D| = \{\, d \in \prod_{n \in \mathbb{N}} D_n \mid \forall n \in \mathbb{N}.\, d_n = f_n(d_{n+1}) \,\}$$

and $d \sqsubseteq_D d'$ iff $\forall n \in \mathbb{N}.\, d_n \sqsubseteq d'_n$ and $q_n : D \to D_n : d \mapsto d_n$, i.e. q_n projects on the n-th component. Notice that $q_n = f_n \circ q_{n+1}$ for all $n \in \mathbb{N}$.

The inverse limit satisfies the universal property that for all sequences $(g_n : E \to D_n \mid n \in \mathbb{N})$ with $g_n = f_n \circ g_{n+1}$ for all $n \in \mathbb{N}$ there exists a unique map $h : E \to D$ with $g_n = q_n \circ h$ for all $n \in \mathbb{N}$.

Proof.   Directed suprema in $D = (|D|, \sqsubseteq_D)$ are computed pointwise which does not lead out of D as the f_n preserve directed suprema. As suprema in D are computed pointwise it readily follows that the q_n preserve them. For $d \in D$ we have $f_n(q_{n+1}(d)) = f_n(d_{n+1}) = d_n = q_n(d)$ where the penultimate equality holds as $d \in |D|$.

For showing that D satisfies the universal property suppose that $(g_n : E \to D_n \mid n \in \mathbb{N})$ with $g_n = f_n \circ g_{n+1}$ for all $n \in \mathbb{N}$. The map $h : E \to D$ with $g_n = q_n \circ h$ for all $n \in \mathbb{N}$ is uniquely determined by this property, i.e. we have $h(y) = (g_n(y))_{n \in \mathbb{N}}$. That this h is Scott continuous follows from the fact that suprema in D are computed pointwise and the assumption that the g_n are all Scott continuous. $\square$

Notice that the universal property of inverse limits determines them uniquely up to isomorphism. If we take inverse limits of sequences consisting of projections we can characterise their inverse limit (up to isomorphism) in a purely local way as follows.

Theorem 9.6   Let $(p_n : D_{n+1} \to D_n \mid n \in \mathbb{N})$ be a sequence of projections. We write e_n for the embeddings associated uniquely with the p_n. Then the q_n of the inverse limit are all projections whose associated embeddings $i_n : D_n \to D$ can be defined explicitly as follows

$$i_n(x)_m = \begin{cases} (e_{m-1} \circ \cdots \circ e_n)(x) & \text{if } n \leq m \\ (p_m \circ \cdots \circ p_{n-1})(x) & \text{otherwise.} \end{cases}$$

Moreover, we have

$$(\ddagger) \qquad \bigsqcup_{n \in \mathbb{N}} i_n \circ q_n = \mathsf{id}_D$$

and this property together with the requirement $q_n = p_n \circ q_{n+1}$ characterises inverse limits up to isomorphism.

Proof. Straightforward computation checks that the i_n are continuous strict maps from D_n to D and that the (i_n, q_n) are embedding/projection pairs. Moreover, one easily sees that $i_{n+1} \circ e_n = i_n$ for all $n \in \mathbb{N}$. Thus, we have

$$i_{n+1} \circ q_{n+1} \sqsupseteq i_{n+1} \circ e_n \circ p_n \circ q_{n+1} = i_n \circ q_n$$

i.e. that the sequence $i_n \circ q_n$ is ascending in $[D \to D]$. As

$$q_n \circ \bigsqcup_{k \in \mathbb{N}} i_k \circ q_k = \bigsqcup_{k \in \mathbb{N}} q_n \circ i_k \circ q_k = \bigsqcup_{k \geq n} q_n \circ i_k \circ q_k = \bigsqcup_{k \geq n} q_n = q_n = q_n \circ \mathsf{id}_D$$

it follows from the universal property of $(q_n)_{n \in \mathbb{N}}$ that $(\ddagger)$ holds.

On the other hand suppose that we have a sequence of embedding/projection pairs (i'_n, q'_n) from D_n to D' with $q'_n = p_n \circ q'_{n+1}$ and $\bigsqcup_{n \in \mathbb{N}} i'_n \circ q'_n = \mathsf{id}_{D'}$. By the universal property of the inverse limit $(q_n)_{n \in \mathbb{N}}$ there is a unique map $\iota : D' \to D$ with $q'_n = q_n \circ \iota$. We show now that this ι is an isomorphism by constructing its (tentative) inverse as

$$\iota^{-1} = \bigsqcup_{n \in \mathbb{N}} i'_n \circ q_n \quad.$$

First notice that from $q'_n = p_n \circ q'_{n+1}$ it follows by Lemma 9.4 that $i'_n = i'_{n+1} \circ e_n$ from which it follows that the sequence $(i'_n \circ q_n)_{n \in \mathbb{N}}$ is ascending and, therefore, the map ι^{-1} is well defined. That $\iota^{-1} \circ \iota = \mathsf{id}_{D'}$ can be seen as follows

$$\iota^{-1} \circ \iota = \left(\bigsqcup_{n \in \mathbb{N}} i'_n \circ q_n \right) \circ \iota = \bigsqcup_{n \in \mathbb{N}} i'_n \circ q_n \circ \iota = \bigsqcup_{n \in \mathbb{N}} i'_n \circ q'_n = \mathsf{id}_{D'} \quad.$$

For showing that $\iota \circ \iota^{-1} = \mathsf{id}_D$ we first observe that

$$(*) \qquad q'_n \circ i'_{n+k} \circ q_{n+k} = q_n$$

for all $k, n \in \mathbb{N}$ which can be seen as follows. For $k{=}0$ claim $(*)$ is obvious as $q'_n \circ i'_n = \mathsf{id}_{D_n}$. Suppose now as induction hypothesis that $q'_n \circ i'_{n+k} \circ q_{n+k} = q_n$ for all $n \in \mathbb{N}$. Now for arbitrary $n \in \mathbb{N}$ it follows from the induction hypothesis that $q'_{n+1} \circ i'_{n+1+k} \circ q_{n+1+k} = q_{n+1}$. Thus, by postcomposition with p_n we get

$$q'_n \circ i'_{n+k+1} \circ q_{n+k+1} = p_n \circ q'_{n+1} \circ i'_{n+1+k} \circ q_{n+1+k} = p_n \circ q_{n+1} = q_n$$

as desired. Now from $(*)$ it follows that

$$q_n \circ \iota \circ \iota^{-1} = q'_n \circ \bigsqcup_{k \in \mathbb{N}} i'_k \circ q_k = \bigsqcup_{k \in \mathbb{N}} q'_n \circ i'_k \circ q_k = \bigsqcup_{k \geq n} q'_n \circ i'_k \circ q_k = \bigsqcup_{k \geq n} q_n = q_n \circ \mathrm{id}_D$$

which by the universal property of $(q_n)_{n \in \mathbb{N}}$ implies $\iota \circ \iota^{-1} = \mathrm{id}_D$.

Thus, we have shown that ι^{-1} is actually the inverse of ι. We have $q_n \circ \iota = q'_n$ by definition of ι and, therefore, also $q_n = q'_n \circ \iota^{-1}$. Thus, the cone $(q'_n)_{n \in \mathbb{N}}$ is isomorphic to the limiting cone $(q_n)_{n \in \mathbb{N}}$ via the isomorphism ι as desired. $\square$

Now we can construct bifree solutions of domain equations $X = F(X, X)$ for arbitrary locally continuous functors $F : \mathcal{C}^{op} \times \mathcal{C} \to \mathcal{C}$.

Theorem 9.7   *Let $F : \mathcal{C}^{op} \times \mathcal{C} \to \mathcal{C}$ be a locally continuous functor. Consider the sequence of embedding/projection pairs (e_n, p_n) from D_n to D_{n+1} defined recursively as follows*

$$D_0 = 1 = \{\bot\} \qquad\qquad D_{n+1} = F(D_n, D_n)$$

$$e_0 = \bot : D_0 \to D_1 \qquad e_{n+1} = F(p_n, e_n) : D_{n+1} \to D_{n+2}$$

$$p_0 = \bot : D_1 \to D_0 \qquad p_{n+1} = F(e_n, p_n) : D_{n+2} \to D_{n+1}$$

and let (i_n, q_n) be the inverse limit for the sequence (e_n, p_n). Then $\alpha = \bigsqcup_{n \in \mathbb{N}} i_{n+1} \circ F(i_n, q_n) : F(D, D) \to D$ is a bifree solution of the domain equation $X = F(X, X)$. The inverse of α is given by $\bigsqcup_{n \in \mathbb{N}} F(q_n, i_n) \circ q_{n+1}$.

Proof.   Straighforward induction (using monotonicity of the morphism part of F) shows that the (e_n, p_n) are actually embedding/projection pairs. By Theorem 9.6 the sequence $i_n \circ q_n$ is ascending and has supremum id_D. From this it follows (again by local monotonicity and continuity of F) that the sequence $F(q_n, i_n) \circ F(i_n, q_n)$ is ascending and has supremum $\mathrm{id}_{F(D,D)}$. Thus, by Theorem 9.6 we know that both $(q_{n+1})_{n \in \mathbb{N}}$ and $(F(i_n, q_n))_{n \in \mathbb{N}}$ are limiting cones for the diagram $(p_{n+1})_{n \in \mathbb{N}}$. From Theorem 9.6 we know that the unique map $\alpha : F(D, D) \to D$ with $q_{n+1} \circ \alpha = F(i_n, q_n)$ is an isomorphism. That $\alpha^{-1} = \bigsqcup_{n \in \mathbb{N}} F(q_n, i_n) \circ q_{n+1}$ follows from inspection of the proof of Theorem 9.6 (namely the construction of ι^{-1} in this proof). Similarly it follows that $\alpha = \bigsqcup_{n \in \mathbb{N}} i_{n+1} \circ F(i_n, q_n)$ since α is the inverse of

α^{-1} which is the mediating arrow from the limiting cone $(q_{n+1})_{n \in \mathbb{N}}$ to the limiting cone $(F(i_n, q_n))_{n \in \mathbb{N}}$.

Thus, for bifreeness it remains to show that every $e : D \circ\!\!\!\to D$ with $e = \delta(e) := \alpha \circ F(e, e) \circ \alpha^{-1}$ is actually equal to id_D. We show that

$$q_n \circ e = q_n \qquad \text{and} \qquad e \circ i_n = i_n$$

for all $n \in \mathbb{N}$ by induction on n from which it follows by the universal property of the limiting cone $(q_n)_{n \in \mathbb{N}}$ that $e = \mathrm{id}_D$. For $n=0$ we have $q_n \circ e = q_n$ as all maps from D to $1 = D_0$ are equal and $e \circ i_n = i_n$ as all strict maps from 1 to D are equal. Suppose as induction hypothesis that $q_n \circ e = q_n$ and $e \circ i_n = i_n$. Then using the induction hypothesis we have

$$q_{n+1} \circ e = q_{n+1} \circ \alpha \circ F(e, e) \circ \alpha^{-1} = F(i_n, q_n) \circ F(e, e) \circ \alpha^{-1}$$

$$= F(e \circ i_n, q_n \circ e) \circ \alpha^{-1} = F(i_n, q_n) \circ \alpha^{-1} =$$

$$= q_{n+1}$$

and

$$e \circ i_{n+1} = \alpha \circ F(e, e) \circ \alpha^{-1} \circ i_{n+1} = \alpha \circ F(e, e) \circ F(q_n, i_n) =$$

$$= \alpha \circ F(q_n \circ e, e \circ i_n) = \alpha \circ F(q_n, i_n) =$$

$$= i_{n+1}$$

proving the induction step. □

Having shown that canonical solutions which are unique up to isomorphism do exist we now give an alternative characterisation of them.

Theorem 9.8 *Let $F : \mathcal{C}^{op} \times \mathcal{C} \to \mathcal{C}$ be a locally continuous functor. Then $\alpha : F(A, A) \to A$ is a bifree solution of the domain equation $X = F(X, X)$ if and only if for all morphisms $f : F(C, B) \to B$ and $g : C \to F(B, C)$ in $\mathcal{C}$ there exist* **unique** *morphisms $h : A \to B$ and $k : C \to A$ in $\mathcal{C}$ making the diagrams*

$$
\begin{array}{ccc}
F(C,B) & \xrightarrow{\ f\ } & B \\
{\scriptstyle F(k,h)}\big\uparrow & & \big\uparrow{\scriptstyle h} \\
F(A,A) & \xleftarrow[\ \alpha^{-1}\ ]{} & A
\end{array}
\qquad
\begin{array}{ccc}
F(B,C) & \xleftarrow{\ g\ } & C \\
{\scriptstyle F(h,k)}\big\downarrow & & \big\downarrow{\scriptstyle k} \\
F(A,A) & \xrightarrow[\ \alpha\ ]{} & A
\end{array}
$$

commute.

Proof.   We first show the implication from right to left. Instantiating B and C by A and f by α and g by α^{-1} we get that from $e = \alpha \circ F(e, e) \circ \alpha^{-1}$ it follows that $e = \mathrm{id}_A$ as both $h = \mathrm{id}_A = k$ and $h = e = k$ make the above diagrams commute and the choice of h and k is unique by assumption.

For the reverse direction assume that α is a bifree solution and that the two diagrams commute. We show that h and k are actually the least (simultaneous) solutions of the equations

$$h = f \circ F(k, h) \circ \alpha^{-1} \qquad \text{and} \qquad k = \alpha \circ F(h, k) \circ g \ .$$

The least solution of this system of equations is given by $\bigsqcup_{n \in \mathbb{N}} h_n$ and $\bigsqcup_{n \in \mathbb{N}} k_n$, respectively, where the h_n and k_n are defined recursively as follows

$$h_0 = \bot : A \to B \qquad\qquad h_{n+1} = f \circ F(k_n, h_n) \circ \alpha^{-1}$$

$$k_0 = \bot : C \to A \qquad\qquad k_{n+1} = \alpha \circ F(h_n, k_n) \circ g.$$

As α is a bifree solution we also have $\mathrm{id}_A = \bigsqcup_{n \in \mathbb{N}} e_n$ where

$$e_0 = \bot : A \to A \qquad\qquad e_{n+1} = \alpha \circ F(e_n, e_n) \circ \alpha^{-1} \ .$$

We show now by induction on n that

$$h_n = h \circ e_n \qquad \text{and} \qquad k_n = e_n \circ k$$

from which it follows that

$$h = h \circ \mathrm{id}_A = h \circ \bigsqcup_{n \in \mathbb{N}} e_n = \bigsqcup_{n \in \mathbb{N}} h \circ e_n = \bigsqcup_{n \in \mathbb{N}} h_n$$

and

$$k = \mathrm{id}_A \circ k = \Big(\bigsqcup_{n \in \mathbb{N}} e_n\Big) \circ k = \bigsqcup_{n \in \mathbb{N}} e_n \circ k = \bigsqcup_{n \in \mathbb{N}} k_n$$

i.e., that the pair (h, k) is the least solution of the equations

$$h = f \circ F(k, h) \circ \alpha^{-1} \qquad \text{and} \qquad k = \alpha \circ F(k, h) \circ g \ .$$

For $n=0$ the claim holds as $\bot = h \circ \bot$ because h is strict and $\bot = \bot \circ k$ holds anyway. Suppose as induction hypothesis that $h_n = h \circ e_n$ and $k_n = e_n \circ k$. Then we have

$$h \circ e_{n+1} = f \circ F(k, h) \circ \alpha^{-1} \circ \alpha \circ F(e_n, e_n) \circ \alpha^{-1} =$$

$$= f \circ F(k, h) \circ F(e_n, e_n) \circ \alpha^{-1} =$$

$$= f \circ F(e_n \circ k, h \circ e_n) \circ \alpha^{-1} = f \circ F(k_n, h_n) \circ \alpha^{-1} =$$

$$= h_{n+1}$$

and

$$e_{n+1} \circ k = \alpha \circ F(e_n, e_n) \circ \alpha^{-1} \circ \alpha \circ F(h, k) \circ g =$$
$$= \alpha \circ F(e_n, e_n) \circ F(h, k) \circ g =$$
$$= \alpha \circ F(h \circ e_n, e_n \circ k) \circ g = \alpha \circ F(k_n, h_n) \circ g =$$

$$= k_{n+1}$$

as desired. $\square$

Notice that in the above proof we have used intrinsically that h is strict. There may arise the question why it is essential to restrict attention to $\mathcal{C}$, i.e. to *strict* continuous maps. This will get clear from the following theorem where one considers the particular case of mixed variant functors induced by locally continuous covariant endofunctors on $\mathcal{C}$ which are also practically most important.

Theorem 9.9 *Let $T : \mathcal{C} \to \mathcal{C}$ be a locally continuous covariant functor. Let $F_T : \mathcal{C}^{op} \times \mathcal{C} \to \mathcal{C}$ be the locally continuous mixed variant functor defined from T by putting $F_T(Y, X) = T(X)$ and $F_T(g, f) = T(f)$. Then an isomorphism $\alpha : T(A) \to A$ is a bifree solution of the domain equation $X = F_T(X, X) = T(X)$ iff one of the following three equivalent conditions is satisfied*

(1) *if $e : A \hookrightarrow A$ with $e = \alpha \circ T(e) \circ \alpha^{-1}$ then $e = \mathrm{id}_A$*
(2) *for every $f : T(B) \hookrightarrow B$ there exists a unique map $h : A \hookrightarrow B$ with $h = f \circ T(h) \circ \alpha^{-1}$, i.e.*

$$
\begin{array}{ccc}
T(B) & \xrightarrow{\ f\ } & B \\
{\scriptstyle T(h)} \uparrow & & \uparrow {\scriptstyle h} \\
T(A) & \xrightarrow[\ \alpha\ ]{} & A
\end{array}
$$

since α is an isomorphism.
(3) *for every $g : B \hookrightarrow T(B)$ there exists a unique map $k : B \hookrightarrow A$ with*

$k = \alpha \circ T(k) \circ g$, *i.e.*

$$
\begin{array}{ccc}
B & \xrightarrow{\ g\ } & T(B) \\
{\scriptstyle k}\downarrow & & \downarrow{\scriptstyle T(k)} \\
A & \xrightarrow[\alpha^{-1}]{} & T(A)
\end{array}
$$

since α is an isomorphism.

Proof.   Obviously, condition (1) is equivalent to α being a bifree solution of $X = F_T(X, X)$. By Theorem 9.8 condition (1) implies conditions (2) and (3). But each of the conditions (2) and (3) entails condition (1) instantiating f and g by α and α^{-1}, respectively.   □

If we allowed in $\mathcal{C}$ also non-strict continuous maps as morphisms then condition (2) of Theorem 9.9 would not be satisfied anymore for bifree solutions of the domain equation $X = X$ for the following reason. Obviously, the isomorphism id_1 where $1 = \{\bot\}$ is a bifree solution for $X = \mathsf{Id}_{\mathcal{C}}(X)$ where $\mathsf{Id}_{\mathcal{C}}$ is the identity functor on $\mathcal{C}$. But now for every domain A and continuous $a : 1 \to A$ we have

$$
\begin{array}{ccc}
A & \xrightarrow{\ \mathrm{id}_A\ } & A \\
{\scriptstyle a}\uparrow & & \uparrow{\scriptstyle a} \\
1 & \xrightarrow[\mathrm{id}_1]{} & 1
\end{array}
$$

but if we do not require a to be strict there are as many a as there are elements of A. Thus, for A with more than one element condition (2) of Theorem 9.9 were violated if we allowed non-strict maps in $\mathcal{C}$.

We conclude this chapter by observing that condition (2) of Theorem 9.9 gives rise to an induction principle for recursive types $A = T(A)$ where $T : \mathcal{C} \to \mathcal{C}$ is locally continuous. Suppose that $P \subseteq A$ is closed under directed suprema and contains $\bot$. Let us write i for the inclusion from P into A. If there exists a map $\pi : T(P) \to P$ with then one easily shows (exercise!) that i is an isomorphism whose inverse is given by the unique $j : A \to P$ with $\pi \circ T(j) = j \circ \alpha$ (using condition (2) of Theorem 9.9). Thus, the map i is surjective and $P = A$. We leave it as an exercise(!) for the inclined reader to show that admissibility of $P \subseteq A$ and $\perp \in P$ are necessary assumptions for this induction principle.

Alas, for general mixed variant locally continuous $F : \mathcal{C}^{op} \times \mathcal{C} \to \mathcal{C}$ such an induction principle is not available. The best we can get is the following. Let $\alpha : F(A, A) \to A$ be a bifree solution and $P \subseteq A$ be closed under directed suprema and $\perp \in P$. Then $a \in P$ iff $\forall n \in \mathbb{N}. \delta^n(\perp)(a) \in P$ where $\delta : \mathcal{C}(A, A) \to \mathcal{C}(A, A) : e \mapsto \alpha \circ F(e, e) \circ \alpha^{-1}$. We leave the verification of this claim to the inclined reader.

Models of Untyped λ-Calculus

Historically, the first domain equation ever considered was $D = [D \to D]$. It was solved by Dana Scott in fall 1969 with the intention of finding mathematical models for untyped λ-calculus where *every* term can be used as a function. Up to that time there were not known any set-theoretic models of the untyped λ-calculus. The reason was that for every set S containing more than one element the set S^S of endofunctions on S has greater cardinality than S itself because of $|S| < |\mathcal{P}(S)| \leq |S^S|$ already known to Cantor. The ingenious idea of Scott was to take instead of a set S a domain D and consider instead of all endofunctions on D just the continuous ones!

Obviously, the bifree solution of $D = [D \to D]$ is trivial, i.e. the trivial domain $1 = \{\perp\}$. To obtain a non-trivial solution D. Scott started with an arbitrary[2] domain R, considered the following sequence of embedding/projection pairs

$$R_0 = R \qquad\qquad R_{n+1} = [R_n \to R_n]$$
$$e_0 = R_0 \to R_1 : r \mapsto \lambda x.r \qquad\qquad e_{n+1} = [p_n \to e_n]$$
$$p_0 : R_1 \to R_0 : f \mapsto f(\perp) \qquad\qquad p_{n+1} = [e_n \to p_n]$$

and showed (as in our Theorem 9.7) that its inverse limit R_∞ is isomorphic to $[R_\infty \to R_\infty]$. Notice, however, that one may show (exercise!) that—as first observed in [Riecke and Sandholm 2002]—the domain R_∞ is isomorphic to $[C \to R]$ where C is the bifree solution of $C = [C \to R] \times C$ rendering the consideration of non-bifree solutions unnecessary.

---

[2] Actually, he used the more common complete lattices instead but, nevertheless, Scott continuous maps as morphisms between them!

A more liberal notion of model for untyped λ-calculus is the following: a domain D such that $[D{\to}D]$ is a retract of D, i.e. there exist continuous maps $s : [D{\to}D] \to D$ and $p : D \to [D{\to}D]$ with $p \circ s = \text{id}_{[D\to D]}$. An example for this is the lattice $\mathcal{P}(\mathbb{N})$ which contains $[\mathcal{P}(\mathbb{N}){\to}\mathcal{P}(\mathbb{N})]$ as a retract in the following way. Let $e : \mathbb{N} \to \mathcal{P}_{\text{fin}}(\mathbb{N})$ be a primitive recursive bijection[3] between natural numbers and finite subsets of them such that the relation $m \in e_n$ is decidable. Then define $p : \mathcal{P}(\mathbb{N}) \to [\mathcal{P}(\mathbb{N}){\to}\mathcal{P}(\mathbb{N})]$ as $p(A)(B) = \{n \in \mathbb{N} \mid \langle m, n \rangle \in A$ and $e_m \subseteq B\}$ and $s(f) = \{\langle m, n \rangle \mid n \in f(e_m)\}$. One easily checks that $p \circ s = \text{id}_{[\mathcal{P}(\mathbb{N})\to\mathcal{P}(\mathbb{N})]}$ and $s \circ p \circ s \circ p = s \circ p \sqsupseteq \text{id}_{\mathcal{P}(\mathbb{N})}$. Notice that for arbitrary infinite sets S one can organize $\mathcal{P}(S)$ into a model of λ-calculus exploiting the fact that $S \cong \mathcal{P}_{\text{fin}}(S) \times S$ for all infinite sets S.

Notice, however, that models of the form $\mathcal{P}(S)$ (usually called *graph models*) do not model η-equality as $[\![\lambda y.\, x(y)]\!]\rho = s(p(\rho(x)))$ whereas $[\![x]\!]\rho = \rho(x)$ and $s \circ p \neq \text{id}_{\mathcal{P}(\mathbb{N})}$. The latter can be seen by considering the set $I = \{\langle 2^n, n \rangle \mid n \in \mathbb{N}\}$ for which we have $s(p(I)) = \{\langle m, n \rangle \mid n \in e_m\}$ strictly bigger than I.

For an in-depth investigation of these models see e.g. the encyclopedic book [Barendregt 1981] by H. Barendregt.

---

[3]e.g. by putting $e(n) = A$ iff $n = \sum_{i \in A} 2^i$

Chapter 10

Characterisation of Fully Abstract Models

The aim of this chapter is to show that extensionally fully abstract models of PCF are unique up to isomorphism. For this purpose we first fix an appropriate notion of model.

Definition 10.1   A *domain-enriched category* $\mathcal{C}$ is given by

(1) a collection $|\mathcal{C}|$ of *objects*
(2) for all $A, B \in |\mathcal{C}|$ a domain $\mathcal{C}(A, B)$ whose least element is denoted by $\bot_{A,B}$
(3) a Scott continuous function $\circ_{A,B,C} : \mathcal{C}(B, C) \times \mathcal{C}(A, B) \to \mathcal{C}(A, C)$ for all objects $A, B, C \in |\mathcal{C}|$
(4) a morphism $\mathrm{id}_A \in \mathcal{C}(A, A)$ for every object $A \in |\mathcal{C}|$

such that

$$\text{(Assoc)} \qquad h \circ (g \circ f) = (h \circ g) \circ f$$

$$\text{(Neutr)} \qquad \mathrm{id} \circ f = f = f \circ \mathrm{id}$$

whenever both sides of the equations are defined.   ◇

Subsequently we often write $f : A \to B$ for $f \in \mathcal{C}(A, B)$.

Definition 10.2   (Λ-category)
A *cartesian closed domain-enriched category* (or *Λ-category*) is a domain-enriched category $\mathcal{C}$ with

(1) an object $1 \in |\mathcal{C}|$ such that for all $A \in |\mathcal{C}|$ there is a unique $f \in \mathcal{C}(A, 1)$, namely $\bot_{A,1}$
(2) for all $A, B \in |\mathcal{C}|$ a distinguished object $A \times B \in |\mathcal{C}|$ and distinguished maps $\pi_1^{A,B} \in \mathcal{C}(A \times B, A)$ and $\pi_2^{A,B} \in \mathcal{C}(A \times B, B)$ such that for all maps $f \in \mathcal{C}(C, A)$ and $g \in \mathcal{C}(C, B)$ there exists a unique map $h \in \mathcal{C}(C, A \times B)$

77

with $\pi_1^{A,B} \circ h = f$ and $\pi_2^{A,B} \circ h = g$ which is denoted by $\langle f, g \rangle$

Notation If $f \in \mathcal{C}(C, A)$ and $g \in \mathcal{C}(D, B)$ then we write $f \times g$ as an abbreviation for $\langle f \circ \pi_1^{C,D}, g \circ \pi_2^{C,D} \rangle$.

(3) for all $A, B \in |\mathcal{C}|$ a distinguished object $[A{\rightarrow}B] \in |\mathcal{C}|$ and a distinguished map $\mathsf{ev}_{A,B} \in \mathcal{C}([A{\rightarrow}B] \times A, B])$ such that for all $f \in \mathcal{C}(C \times A, B)$ there is a unique map $g \in \mathcal{C}(C, [A{\rightarrow}B])$ with $\mathsf{ev}_{A,B} \circ (g \times \mathrm{id}_A) = f$ for which we write $\Lambda_{A,B,C}(f)$ or simply $\Lambda(f)$

such that

(i) $\langle f, g \rangle \sqsubseteq \langle f', g' \rangle$ whenever $f \sqsubseteq f'$ and $g \sqsubseteq g'$
(ii) $\Lambda(f) \sqsubseteq \Lambda(g)$ whenever $f \sqsubseteq g$
(iii) $\bot \circ f = \bot$
(iv) $\mathsf{ev} \circ \langle \bot, a \rangle = \bot$

where the last two equations are required to hold whenever both sides of the equation are defined.   ◇

It follows from the above axioms that for all objects A, B and C the mappings

$$\langle -, - \rangle : \mathcal{C}(C, A) \times \mathcal{C}(C, B) \to \mathcal{C}(C, A \times B)$$

and

$$\Lambda : \mathcal{C}(C \times A, B) \to \mathcal{C}(C, [A{\rightarrow}B])$$

are order isomorphisms and thus continuous. Their inverses are given by the assignments $h \mapsto (\pi_1 \circ h, \pi_2 \circ h)$ and $g \mapsto \mathsf{ev} \circ (g \times \mathrm{id}_A)$, respectively.

Next we show that in Λ-categories one has least fixpoint operators.

Lemma 10.3   *Let $\mathcal{C}$ be a Λ-category and $A \in |\mathcal{C}|$. Then there exists a least morphism* $\mathsf{fix}_A \in \mathcal{C}([A{\rightarrow}A], A)$ *with* $\mathsf{ev} \circ \langle \mathrm{id}_{[A{\rightarrow}A]}, \mathsf{fix}_A \rangle = \mathsf{fix}_A$.

Proof.   For $A \in |\mathcal{C}|$ the function

$$F_A : \mathcal{C}([A{\rightarrow}A], A) \to \mathcal{C}([A{\rightarrow}A], A) : h \mapsto \mathsf{ev} \circ \langle \mathrm{id}_{[A{\rightarrow}A]}, h \rangle$$

is Scott continuous due to the properties required for a Λ-category. Thus, the mapping F_A has a least fixpoint $\mathsf{fix}_A = \bigsqcup_{n \in \mathbb{N}} F_A^n(\bot_{[A{\rightarrow}A], A})$.   □

Using the fix of the previous lemma one can construct least fixpoints in the following way.

Theorem 10.4   *Let* C *be a* Λ-*category and* $f \in C(C, [A{\to}A])$. *Then* $\mathsf{fix}_A \circ f$
is the least morphism $a \in C(C, A)$ *satisfying*

$$a = \mathsf{ev} \circ \langle f, a \rangle$$

where fix_A *is as in Lemma 10.3.*

Proof.   Obviously, for $f : C \to [A{\to}A]$ we have

$$\mathsf{fix}_A \circ f = \mathsf{ev} \circ \langle \mathsf{id}_{[A \to A]}, \mathsf{fix}_A \rangle \circ f = \mathsf{ev} \circ \langle \mathsf{id}_{[A \to A]} \circ f, \mathsf{fix}_A \circ f \rangle = \mathsf{ev} \circ \langle f, \mathsf{fix}_A \circ f \rangle$$

due to the defining equation for fix_A. Now if $a : C \to A$ satisfies the
inequality $\mathsf{ev} \circ \langle f, a \rangle \sqsubseteq a$ then it follows by fixpoint induction that $\mathsf{fix}_A \circ f \sqsubseteq a$
because

$$\bot_{[A \to A], A} \circ f = \bot_{C, A} \sqsubseteq a$$

and

$$F_A(h) \circ f = \mathsf{ev} \circ \langle \mathsf{id}_{[A \to A]}, h \rangle \circ f = \mathsf{ev} \circ \langle f, h \circ f \rangle \sqsubseteq \mathsf{ev} \circ \langle f, a \rangle \sqsubseteq a$$

whenever $h \circ f \sqsubseteq a$.   □

As $C(A, A) \cong C(1 \times A, A) \cong C(1, [A{\to}A])$ by $f \mapsto f \circ \pi_2^{1,A} \mapsto$
$\Lambda(f \circ \pi_2^{1,A}) =: \ulcorner f \urcorner$ one easily shows (exercise!) that $\mathsf{fix}_A \circ \ulcorner f \urcorner : 1 \to A$
is the least $a : 1 \to A$ with $a = f \circ a$.

From the above considerations it appears that a Λ-category is endowed
with enough structure to interpret typed λ-calculus with fixpoint operators
at all types. For interpreting full PCF one just needs an object N together
with appropriate morphisms $\mathsf{zero} : 1 \to N$, $\mathsf{succ}, \mathsf{pred} : N \to N$ and $\mathsf{ifz} :$
$N \times N \times N \to N$ for interpreting base type **nat** and the basic operations on
it. In order to formulate the necessary requirements we need some notation
introduced in the next definition.

Definition 10.5   Let C be a Λ-category. For $A \in |C|$ we write $\Gamma(A)$ as an
abbreviation for the domain $C(1, A)$ and for morphisms $f : A \to B$ in C we
write $\Gamma(f)$ for the continuous map $\Gamma(A) \to \Gamma(B) : a \mapsto f \circ a$.   ◇

The elements of $\Gamma(A)$ are often referred to as "global elements of A".

Definition 10.6   (N-structure)
An N-*structure* or *natural numbers structure* in a Λ-category is given by

- an object $N \in |C|$
- $\mathsf{zero} \in C(1, N)$

- succ, pred : $N \to N$
- ifz : $N \times N \times N \to N$

such that the map $i_N : \mathbb{N}_\perp \to \Gamma(N)$ sending $\perp$ to $\perp_{1,N}$ and $n \in \mathbb{N}$ to $\mathsf{succ}^n \circ \mathsf{zero}$ is an isomorphism of domains and the following conditions are satisfied

(1) $\mathsf{pred} \circ \mathsf{zero} = \mathsf{zero}$ and $\mathsf{pred} \circ \mathsf{succ} = \mathsf{id}_N$

(2) $\mathsf{ifz} \circ \langle \mathsf{zero} \circ !_{N \times N}, \pi_1^{N,N}, \pi_2^{N,N} \rangle = \pi_1^{N,N}$ and

$\mathsf{ifz} \circ \langle \mathsf{succ}^{n+1} \circ \mathsf{zero} \circ !_{N \times N}, \pi_1^{N,N}, \pi_2^{N,N} \rangle = \pi_2^{N,N}$ and

$\mathsf{ifz} \circ \langle \perp_{N \times N, N}, \pi_1^{N,N}, \pi_2^{N,N} \rangle = \perp_{N \times N, N}.$ $\diamond$

We are particularly interested in Λ-categories where equality and order on morphisms is determined by their behaviour on global elements.

Definition 10.7 ((order) extensional)
A Λ-category $\mathcal{C}$ is called *extensional* iff $\Gamma(f) = \Gamma(g)$ implies $f = g$ for all morphisms $f, g : A \to B$ in $\mathcal{C}$ and it is called *order extensional* iff $\Gamma(f) \sqsubseteq \Gamma(g)$ implies $f \sqsubseteq g$ for all morphisms $f, g : A \to B$ in $\mathcal{C}$. $\diamond$

The notions of extensionality and order extensionality can be explicitated as follows: the Λ-category $\mathcal{C}$ is *extensional* iff

$$\forall a \in \mathcal{C}(1, A).\; f \circ a = g \circ b \in \mathcal{C}(1, B) \Rightarrow f = g \in \mathcal{C}(A, B)$$

and it is *order extensional* iff

$$\forall a \in \mathcal{C}(1, A).\; f \circ a \sqsubseteq g \circ b \in \mathcal{C}(1, B) \Rightarrow f \sqsubseteq g \in \mathcal{C}(A, B) .$$

In an extensional Λ-category the continuous function

$$\mathsf{App} : \Gamma(B^A) \times \Gamma(A) \to \Gamma(B) : (f, a) \mapsto \mathsf{ev} \circ \langle f, a \rangle$$

induces by functional abstraction the function

$$\Lambda(\mathsf{App}) : \Gamma(B^A) \to \Gamma(B)^{\Gamma(B)}$$

which is injective iff the model is extensional and which reflects the order iff the model is order extensional in which latter case $\Gamma(B^A)$ appears as a subposet of $\Gamma(B)^{\Gamma(A)}$ via $\Lambda(\mathsf{App})$.

Now we define a notion of PCF model appropriate for our later characterisation of full abstraction.

Definition 10.8 (PCF model)
A PCF *model* is a Λ-category $\mathcal{C}$ together with an N-structure in $\mathcal{C}$. Such a model is called *(order) extensional* iff $\mathcal{C}$ is (order) extensional.

In extensional PCF models we make no notational distinction between A and $\Gamma(A)$ and between f and $\Gamma(f)$ since no information is lost when restricting attention to global elements. $\diamond$

One can interpret PCF in arbitrary PCF models and not only in the Scott model as the semantic equations of Definition 3.14 do still make sense when reformulated in an "element-free" way (exercise!). One easily shows that the evaluation relation preserves semantic equality w.r.t. such models, i.e. if $\Gamma \vdash M : \sigma$ and $M{\Downarrow}V$ then $[\![\Gamma \vdash M]\!] = [\![\Gamma \vdash V]\!]$. The proof of this correctness property can be copied almost verbatim from the proof for the case of the Scott model.

Notice also that for PCF models one may prove computational adequacy essentially in the same way as in Chapter 4.

As in Chapter 8 using the interpretations of the PCF terms ψ_n^σ one can show (see [Stoughton 1990]) that in *extensional* PCF models the $[\![\sigma]\!]$ are all SFP domains (as for this purpose one does not need that the order on the $[\![\sigma{\to}\tau]\!]$ is pointwise). In case of order extensionality one can even show (see [Stoughton 1990]) that all $[\![\sigma]\!]$ are Scott domains because in this case infima in functions spaces are pointwise and thus PCF definable (see Chapter 8).

Following [Stoughton 1990] (Theorem 5.7) we show now that in extensional equationally fully abstract PCF models all compact elements of PCF types are PCF definable.

Lemma 10.9 *In equationally fully abstract* PCF *models all compact elements of* PCF *types arise as interpretations of closed* PCF *terms.*

Proof. For the sake of deriving a contradiction suppose that not all compact elements of PCF types are PCF definable. Then there is a minimal such type $\sigma = \sigma_1{\to}\ldots{\to}\sigma_n{\to}\iota$ where the compact elements of the $[\![\sigma_i]\!]$ are all PCF definable but some $e = [\![\psi_k^\sigma]\!](e) \in [\![\sigma]\!]$ is *not* PCF definable.

Let $\leq$ denote the pointwise order of $[\![\sigma]\!]$, i.e. $f \leq g$ iff $fa_1\ldots a_n \sqsubseteq ga_1\ldots a_n$ for all $a_i \in [\![\sigma_i]\!]$. As f and g are continuous and the $[\![\sigma_i]\!]$ are SFP domains we have $f \leq g$ iff $fe_1\ldots e_n \sqsubseteq ge_1\ldots e_n$ for all $e_i \in \mathcal{K}([\![\sigma_i]\!])$. It is easy to see (exercise!) that $f \sqcap g := [\![\mathsf{glb}^\sigma]\!]fg$ (where glb is defined as in Lemma 8.1) is the infimum of f and g w.r.t $\leq$.

Let K be the set of PCF definable elements in the image of $[\![\psi_k^\sigma]\!]$ and $K = K^+ \dot\cup K^-$ where $K^+ = \{d \in K \mid e \leq d\}$.[1] As K is finite the elements of K^+ and K^- can be enumerated, i.e. $K^+ = \{d_0, \ldots, d_{p-1}\}$ and $K^- = \{c_0, \ldots, c_{q-1}\}$. For every $i < q$ let b_j^i be compact elements in $[\![\sigma_j]\!]$ for $j = 1, \ldots, n$ such that

$$eb_1^i \ldots b_n^i \not\sqsubseteq c_i b_1^i \ldots b_n^i$$

which must exists as otherwise $e \leq c_i$ contradicting $c_i \notin K^+$. By minimality of σ all the b_j^i are PCF definable by some closed PCF term B_j^i.

Now we distinguish two cases.

Case 1 : K^+ is nonempty
As all d_i are PCF definable their infimum $d := d_0 \sqcap \cdots \sqcap d_{p-1}$ (w.r.t. $\leq$) is PCF definable, too. As by assumption e is not PCF definable we have $e \not\leq d$. Thus, there exist compact elements $a_1, \ldots, a_n$ with

$$ea_1 \ldots a_n = \bot \qquad \text{and} \qquad da_1 \ldots a_n \neq \bot$$

as otherwise $e \leq d$.

For $x \in [\![\iota]\!]$ let $[x \backslash 0]$ be a PCF term denoting the function $f_x : D_\iota \to D_\iota$ with $f_x(y) = 0$ if $x \sqsubseteq y$ and $f_x(y) = \bot$ otherwise. For $i < q$ consider the closed PCF terms

$$M_i \equiv \lambda f{:}\sigma.\, [eb_1^i \ldots b_n^i \backslash 0](\psi_k^\sigma f B_1^i \ldots B_n^i)$$

and furthermore the closed PCF terms

$$N_1 \equiv \lambda f{:}\sigma.\, [ea_1 \ldots a_n \backslash 0](\psi_k^\sigma f A_1 \ldots A_n)$$
$$N_2 \equiv \lambda f{:}\sigma.\, [da_1 \ldots a_n \backslash 0](\psi_k^\sigma f A_1 \ldots A_n)$$

where the A_i are closed PCF terms with $[\![A_i]\!] = a_i$ (which exist since compact elements of the $[\![\sigma_i]\!]$ are PCF definable by minimality of σ).

Now for $M \equiv M_0 \sqcap \cdots \sqcap M_{q-1}$ we have $[\![M]\!](e) = 0$ since $[\![M_i]\!](e) = 0$ for $i < q$. As $e = [\![\psi_k^\sigma]\!](e)$ we have $[\![N_1]\!](e) = 0$ and $[\![N_1]\!](e) = \bot$ because $da_1 \ldots a_n \not\sqsubseteq ea_1 \ldots a_n$. Thus, we have $[\![M \sqcap N_1]\!](e) = 0$ whereas $[\![M \sqcap N_2]\!](e) = \bot$ from which it follows that the denotations of $M \sqcap N_1$ and $M \sqcap N_2$ are different. But for arbitrary closed terms N of type σ we have

$$[\![M \sqcap N_1]\!]([\![N]\!]) = 0 = [\![M \sqcap N_2]\!]([\![N]\!]) \qquad \text{if } [\![\psi_k^\sigma N]\!] \in K^+$$

---

[1]We use the symbol $\dot\cup$ to denote disjoint union, i.e. $M = M_1 \dot\cup M_2$ iff $M = M_1 \cup M_2$ and $M_1 \cap M_2 = \emptyset$.

as then $ea_1 \ldots a_n \sqsubseteq [\![\psi_k^\sigma N]\!] a_1 \ldots a_n$ and

$$[\![M \sqcap N_1]\!]([\![N]\!]) = \bot = [\![M \sqcap N_2]\!]([\![N]\!]) \qquad \text{if } [\![\psi_k^\sigma N]\!] \in K^-$$

as then for $i < q$ with $c_i = [\![\psi_k^\sigma N]\!]$ we have $[\![M_i]\!]([\![N]\!]) = \bot$ because $eb_1^i \ldots b_n^i \not\sqsubseteq c_i b_1^i \ldots b_n^i$. Thus, by Milner's Context Lemma the terms $M \sqcap N_1$ and $M \sqcap N_2$ are observationally equal although their denotations are different in contradiction with the assumption that the model under consideration is equationally fully abstract.

Case 2 : K^+ is empty
For the M as defined above we have $[\![M]\!](e) = 0$ whereas $[\![\Omega_\sigma]\!](e) = \bot$ and thus $[\![M]\!] \neq [\![\Omega_\sigma]\!]$. For arbitrary closed terms N of type σ we have $[\![\psi_k^\sigma N]\!] \in K^-$ (as $K = K^-$) and, therefore,

$$[\![M]\!]([\![N]\!]) = \bot = [\![\Omega_\sigma]\!]([\![N]\!])$$

since for $i < q$ with $c_i = [\![\psi_k^\sigma N]\!]$ we have $[\![M_i]\!]([\![N]\!]) = \bot$ because $eb_1^i \ldots b_n^i \not\sqsubseteq c_i b_1^i \ldots b_n^i$. Thus, by Milner's Context Lemma the terms M and Ω_σ are observationally equal although their denotations are different in contradiction with the assumption that the model under consideration is equationally fully abstract. $\qquad \square$

This lemma is crucial for proving the following characterisation of equational full abstraction for extensional PCF models.

Theorem 10.10 (Characterisation of Equational Full Abstraction)
An extensional PCF model is equationally fully abstract iff all compact elements of PCF types are PCF definable.

Proof. Lemma 10.9 says that all compact elements in an extensional equationally fully abstract PCF model are PCF definable.

For the reverse direction suppose we are given an extensional PCF model where all compact elements of PCF types are PCF definable. Suppose M and N are observationally equal closed PCF terms. Then $[\![M]\!]$ and $[\![N]\!]$ are equal on all PCF definable arguments and, therefore, on all compact arguments since these are all assumed to be PCF definable. Thus, as $[\![M]\!]$ and $[\![N]\!]$ are continuous maps and all interpretations of PCF types are SFP domains it follows that $[\![M]\!] = [\![N]\!]$ as desired. $\qquad \square$

Theorem 10.11 (Characterization of Full Abstraction)
An extensional PCF model is fully abstract iff it is order extensional and all compact elements of PCF types are PCF definable.

Proof. Suppose an extensional PCF model is fully abstract. Then it is in particular also equationally fully abstract from which it follows by Theorem 10.10 that all compact elements of PCF types are PCF definable. For showing that the model is also order extensional suppose $f, g \in [\![\sigma \to \tau]\!]$ with $f(a) \sqsubseteq g(a)$ for all $a \in [\![\sigma]\!]$. For $n \in \mathbb{N}$ let $f_n = [\![\psi_n^{\sigma \to \tau}]\!](f)$ and $g_n = [\![\psi_n^{\sigma \to \tau}]\!](g)$, respectively. Obviously, for all $a \in [\![\sigma]\!]$ we have $f_n(a) \sqsubseteq g_n(a)$. Thus, as f_n and g_n are compact and thus PCF definable it follows that $f_n \sqsubseteq g_n$ since the model is fully abstract and $f_n(a) \sqsubseteq g_n(a)$ for all PCF definable $a \in [\![\sigma]\!]$. As $f_n \sqsubseteq g_n$ for al $n \in \mathbb{N}$ it follows that $f = \bigsqcup_{n \in \mathbb{N}} f_n \sqsubseteq \bigsqcup_{n \in \mathbb{N}} g_n = g$.

For the reverse direction suppose we are given an order extensional PCF model where all compact elements of PCF types are PCF definable. First recall that an (order) extensional PCF model is computationally adequate. Now suppose that $M \precsim_\sigma N$. If $\sigma = \iota$ then by computational adequacy $[\![M]\!] \sqsubseteq [\![N]\!]$ because all elements of $[\![\iota]\!]$ which are not numerals are equal to $\bot$. If σ is a functional type then by computational adequacy we have $[\![M]\!](\vec{d}) \sqsubseteq [\![N]\!](\vec{d})$ for all PCF definable arguments $\vec{d}$. Thus, as all compact elements in the model are PCF definable, $[\![M]\!]$ and $[\![N]\!]$ are continuous functions and all domains in the model are SFP domains it follows by order extensionality that $[\![M]\!] \sqsubseteq [\![N]\!]$ as desired.   □

See [Stoughton 1990] for an example of an extensional equationally fully abstract PCF model which, however, is *not* fully abstract, i.e. where for some closed terms M and N it holds that $M \precsim N$ although $[\![M]\!] \not\sqsubseteq [\![N]\!]$. Thus, extensional equationally fully abstract PCF models are not unique up to isomorphism. However, as a consequence of Theorem 10.11, we get the following uniqueness result for extensional fully abstract models.

Theorem 10.12   (Uniqueness of Extensional Fully Abstract Models) *Extensional fully abstract models of* PCF *are unique up to isomorphism. In particular, each $[\![\sigma]\!]$ is isomorphic to the ideal completion of the preorder $(\mathsf{FT}_\sigma, \precsim_\sigma)$ where FT_σ is the set of all $\psi_n^\sigma(M)$ with $M \in \mathsf{Prg}_\sigma$ and $n \in \mathbb{N}$.*

Proof. From Lemma 10.9 it follows that an element of $[\![\sigma]\!]$ is compact iff it can be denoted by some term of the form $\psi_n^\sigma(M)$. As in a fully abstract model $M \precsim_\sigma N$ iff $[\![M]\!] \sqsubseteq [\![N]\!]$ it follows that the subposet $\mathcal{K}([\![\sigma]\!])$ of $[\![\sigma]\!]$ is isomorphic to FT_σ modulo[2] $\precsim_\sigma$. Thus, as $[\![\sigma]\!]$ is an SFP domain it follows that $[\![\sigma]\!] \cong \mathsf{Idl}(\mathsf{FT}_\sigma, \precsim_\sigma)$. Accordingly, all extensional fully abstract models of PCF are isomorphic.   □

---

[2]meaning that $M, N \in \mathsf{FT}_\sigma$ get identified iff $M \precsim_\sigma N$ and $N \precsim_\sigma M$

This theorem allows one to construct the extensional fully abstract model of PCF as the ideal completion of some order-enriched many-sorted algebra. This construction was first performed by R. Milner in [Milner 1977] already back in 1977 and, therefore, the extensional fully abstract model of PCF is commonly called the *Milner model*.

However, people were not satisfied by Milner's model construction because it is *not at all syntax-free* since the relation $\lesssim$ is defined in terms of the operational semantics of PCF. Moreover, in Milner's construction the elements of function types are ideals of equivalence classes of terms instead of—as one might hope—in terms of continuous functions preserving some additional structure common to all Scott domains in the Milner model.

In Chapter 11 we will identify such a structure. We will define so-called "sequential domains" which are domains together with a huge bunch of relations of finite arity and require morphisms between sequential domains to be continuous maps which preserve all the relational structure. In the subsequent Chapter 12 we will then show that the category $\mathcal{S}$ of sequential domains hosts the Milner model of PCF.

Notice, however, that in [Normann 2006] Dag Normann has shown that already at type level 3 the Milner model contains functionals which cannot be computed by a sequential strategy. This means that already at type level 3 there are functionals which are not sequential but can be obtained as the pointwise supremum of an increasing (w.r.t. the pointwise order) chain of PCF definable functionals.[3]

---

[3]This also gives a negative answer to the old question whether the fully abstract models considered in *Game Semantics* (see [Hyland and Ong 2000; Abramsky et.al. 2000]) are cpo-enriched.

Chapter 11

Sequential Domains as a Model of PCF

In this chapter, based on previous work by K. Sieber, P. O'Hearn, J. Riecke and A. Sandholm (see [Sieber 1992; O'Hearn and Riecke 1995; Riecke and Sandholm 2002]), we define a category $\mathcal{S}$ of *sequential domains* which hosts a model of PCF which in the next chapter will be shown to be fully abstract.

Definition 11.1   (partial partition)
Let w be a finite subset of $\mathbb{N}$. A *partial partition* of w is a subset P of $\mathcal{P}(w)$ such that

(1) every element u of P is non-empty and
(2) all $u, v \in P$ are either equal or disjoint, i.e.
$$u, v \in P \Rightarrow u = v \vee u \cap v = \emptyset.$$

We write $\mathsf{pPart}(w)$ for the set of all partial partitions of w. $\Diamond$

Notice that every $P \in \mathsf{pPart}(w)$ is a partition of the set $\bigcup P$.

Definition 11.2   (ssp)
A *structural system of partitions (ssp)* on a finite subset w of $\mathbb{N}$ is a subset $S \subseteq \mathsf{pPart}(w)$ such that

(SSP1) $\{w\} \in S$

(SSP2)   whenever $u \in P \in S$ then $P\backslash\{u\} \in S$

(SSP3)   whenever $S \in P$ and $u, v \in P$ then $(P\backslash\{u, v\}) \cup \{u \cup v\} \in P$

(SSP4)   whenever $P, Q \in S$ and $u \in P$
then $(P\backslash\{u\}) \cup (\{u \cap v \mid v \in Q\}\backslash\{\emptyset\}) \in S$.

We say that (w, S) is *structural system of partitions (ssp)* iff S is a structural system of partitions on $w \in \mathcal{P}_{\mathsf{fin}}(\mathbb{N})$. A *homomorphism from* (w_1, S_1) *to*

(w_2, S_2) is a function $h : w_1 \to w_2$ such that

$$\{h^{-1}[u] \mid u \in P\} \backslash \{\emptyset\} \in S_1$$

for all $P \in S_2$. We write $h : (w_1, S_1) \to (w_2, S_2)$ if h is a homomorphism from (w_1, S_1) to (w_2, S_2). Obviously, ssp-homorphisms are closed under composition and $\mathrm{id}_w : (w, S) \to (w, S)$ for all ssp's (w, S). Thus, structural systems of partitions and their homomorphisms form a category which we denote by **SSP**.   ◊

The intuition behind ssp's (w, S) is that they are "finite data types" (with underlying set w) together with the collection S of those partial partitions that are induced by partial maps

$$f : w \rightharpoonup \{1, \ldots, n\}$$

arising from functional programs on this data type to some "enumeration type" $\{1, \ldots, n\}$. The requirements (SSP1)–(SSP4) reflect some obvious closure properties of such partitions induced by functional programs:

(SSP1) says that we always have the trivial partition as induced by constant total maps which certainly arise from programs.

(SSP2) says that from any $P \in S$ one may remove any of its elements corresponding to the fact that if $f : w \rightharpoonup \{1, \ldots, n\}$ arises from a program then the function f' with $f'(k)\uparrow$ if $f(k) = i$ and $f'(k) = f(k)$ otherwise arises from a program, too.

(SSP3) says that if from some $P \in S$ we remove $u, v \in P$ and instead add $u \cup v$ then the resulting set of equivalence classes is also in S corresponding to the fact that if $f : w \rightharpoonup \{1, \ldots, n\}$ arises from a program then f' with $f'(k) = i$ if $f(k) = i$ or $f(k) = j$ and $f'(k) = f(k)$ otherwise arises from a program, too.

(SSP4) says that if for some $P \in S$ we remove $u \in P$ and replace it for some $Q \in S$ by all non-empty intersections $u \cap v$ with $v \in Q$ corresponding to the fact that if $f : w \rightharpoonup \{1, \ldots, n\}$ and $g : w \rightharpoonup \{1, \ldots, m\}$ arise from programs then for every i with $1 \leq i \leq n$ the function $h : w \rightharpoonup \{1, \ldots, n+m-1\}$ with

$$h(k) = \begin{cases} f(k) & \text{if } f(k) < i \\ i{-}1{+}g(k) & \text{if } f(k) = i \text{ and } g(k)\downarrow \\ f(k){+}m & \text{if } f(k) > i \\ \uparrow & \text{otherwise} \end{cases}$$

arises from a program, too.

We now define what is a logical relation of arity (w, S) on a domain A.

Definition 11.3   (logical relation of arity (w, S))
Let (w, S) be a ssp and A a domain. A *logical relation on A of arity (w, S)*
is a subset $R \subseteq A^w$ such that

(R1) R is closed under directed suprema (taken in A^w)

(R2) R contains all constant maps from w to A,
       i.e. $\delta(a) := \underline{\lambda} i \in w.a \in R$ for all $a \in A$

(R3)   whenever $f \in R$ and $\{u\} \in S$ then $f \upharpoonright u \in R$ where

$$(f \upharpoonright u)(i) = \begin{cases} f(i) & \text{if } i \in u \\ \bot & \text{otherwise} \end{cases}$$

(R4) if $f \in A^w$ and $P \in S$ with $f \upharpoonright u \in R$ for all $u \in P$ then $f \upharpoonright \bigcup P \in R$. $\Diamond$

Condition (R3) expresses stability under restriction and condition (R4)
says that one may " glue elements of R w.r.t. some $P \in S$ ".

Definition 11.4   (Kripke logical relation)
A *(varying) arity* is a (in general non-full) subcategory of **SSP**.

For a varying arity $\mathcal{A}$ and a domain A a *Kripke logical relation on A
of arity $\mathcal{A}$* is a function R assigning to every $(w, S) \in |\mathcal{A}|$ a logical relation
$R^{(w,S)}$ on A of arity (w, S) in such a way that $f \circ h \in R^{(w',S')}$ whenever
$f \in R^{(w,S)}$ and $h : (w', S') \to (w, S)$ is an arrow in $\mathcal{A}$. $\Diamond$

Now we have collected enough notions to define the category $\mathcal{S}$ of sequential domains.

Definition 11.5   (Category $\mathcal{S}$ of Sequential Domains)
A *sequential domain A* consists of a domain $|A|$ together with a mapping
$R(A)$ which assigns to every varying arity $\mathcal{A}$ a Kripke logical relation $R(A)_{\mathcal{A}}$
on $|A|$ of arity $\mathcal{A}$.

Sequential domains form the objects of a category $\mathcal{S}$ whose morphisms
are defined as follows: an *$\mathcal{S}$-morphism from A to B* is a Scott continuous
function $f : |A| \to |B|$ such that for all varying arities $\mathcal{A}$ it holds that

$$a \in R(A)_{\mathcal{A}}^{(w,S)} \implies f \circ a \in R(B)_{\mathcal{A}}^{(w,S)}$$

for all $(w, S) \in |\mathcal{A}|$. Composition of morphisms and identity maps are
inherited from the category of domains and continuous maps. $\Diamond$

The category of domains and Scott continuous maps is contained in $\mathcal{S}$ as the full subcategory on those objects $A = (|A|, R(A))$ where $R(A)_{\mathcal{A}}^{(w,S)} = A^w$ for all varying arities $\mathcal{A}$ and all objects $(w, S) \in \mathcal{A}$ because any Scott continuous function between such objects necessarily preserves all relations as required by the above definition of an $\mathcal{S}$-morphism.

The following lemma gives a non-trivial example of a sequential domain that will later serve as the interpretation of **nat** in the fully abstract model.

Lemma 11.6 *Let $|N| = \mathbb{N}_\perp$ and for all arities $\mathcal{A}$ and $(w, S) \in |\mathcal{A}|$ let*

$$f \in R(N)_{\mathcal{A}}^{(w,S)} \quad \Longleftrightarrow \quad \{f^{-1}[\{n\}] \mid n \in \mathbb{N}\} \setminus \{\emptyset\} \in S$$

for $f : w \to \mathbb{N}_\perp$. Then $N = (|N|, R(N))$ is an object of $\mathcal{S}$.

Proof. Straightforward, but lengthy exercise! □

Next we show that the category $\mathcal{S}$ is cartesian closed.

Theorem 11.7 *The category $\mathcal{S}$ is cartesian closed, i.e. has finite products and exponential objects.*

Proof. The terminal object 1 is given by $|1| = \{\perp\}$ and $R(1)_{\mathcal{A}}^{(w,S)} = \{\perp\}^w$. One readily checks that for every $A \in |\mathcal{S}|$ the constant map from $|A|$ to $\{\perp\}$ is the unique $\mathcal{S}$-morphism from A to 1.

Let A and B be arbitrary sequential domains for which we will give the construction of $A \times B$ and $[A \to B]$ next.

The underlying domain of $A \times B$ is the cartesian product of the underlying domains of A and B, respectively, i.e. $|A \times B| = |A| \times |B|$, and

$$f \in R(A \times B)_{\mathcal{A}}^{(w,S)} \quad \text{iff} \quad \pi_1 \circ f \in R(A)_{\mathcal{A}}^{(w,S)} \text{ and } \pi_2 \circ f \in R(B)_{\mathcal{A}}^{(w,S)}$$

where π_1 and π_2 are first and second projection, respectively. Thus, the continuous maps $\pi_1 : A \times B \to A$ and $\pi_2 : A \times B \to B$ preserve all relations and, accordingly, are $\mathcal{S}$-morphisms. We leave it as a simple exercise(!) to show that for $\mathcal{S}$-morphisms $f : C \to A$ and $g : C \to B$ their target tupling $\langle f, g \rangle$ is an $\mathcal{S}$-morphism from C to $A \times B$.

The underlying domain of $[A \to B]$ is $\mathcal{S}(A, B)$, the set of $\mathcal{S}$-morphisms from A to B, under the pointwise ordering. One readily checks (using intrinsically condition (R1) from Definition 11.3) that $\mathcal{S}(A, B)$ is closed under directed suprema taken in $[|A| \to |B|]$. Due to condition (R2) of Definition 11.3 every constant map from $|A|$ to $|B|$ is in $\mathcal{S}(A, B)$ and thus the map $\lambda x : |A|.\perp_{|B|}$ is the least element of $\mathcal{S}(A, B)$. For variable arities

$\mathcal{A}$ and $(w, S) \in |\mathcal{A}|$ we define the relation $R([A{\to}B])_{\mathcal{A}}^{(w,S)}$ as follows: $f \in R([A{\to}B])_{\mathcal{A}}^{(w,S)}$ iff

$$\mathsf{ev} \circ \langle f{\circ}h, a \rangle = \underline{\lambda}j{\in}w'.f(h(j))(a(j)) \in R(B)_{\mathcal{A}}^{(w',S')}$$

for all $h : (w', S') \to (w, S)$ in $\mathcal{A}$ and $a \in R(A)_{\mathcal{A}}^{(w',S')}$. If all $R([A{\to}B])_{\mathcal{A}}^{(w,S)}$ are logical relations in the sense of Definition 11.3 then it is obvious that the $R([A{\to}B])_{\mathcal{A}}$ are actually Kripke logical relations of arity $\mathcal{A}$ (because of the quantification over all $h : (w', S') \to (w, S)$ in $\mathcal{A}$).

That $R([A{\to}B])_{\mathcal{A}}^{(w,S)}$ satisfies conditions (R1) and (R2) of Definition 11.3 is left as an easy exercise(!).

For verifying condition (R3) suppose that $f \in R([A{\to}B])_{\mathcal{A}}^{(w,S)}$ and $\{u\} \in S$. We have to show that $f{\upharpoonright}u \in R([A{\to}B])_{\mathcal{A}}^{(w,S)}$, too. For that purpose suppose that $h : (w', S') \to (w, S)$ is in $\mathcal{A}$ and $a \in R(A)_{\mathcal{A}}^{(w',S')}$. Due to the assumption $f \in R([A{\to}B])_{\mathcal{A}}^{(w,S)}$ we know that $\mathsf{ev} \circ \langle f{\circ}h, a \rangle \in R(B)_{\mathcal{A}}^{(w',S')}$ and, therefore, also $\mathsf{ev}{\circ}\langle f{\circ}h, a \rangle \upharpoonright h^{-1}[u] \in R(B)_{\mathcal{A}}^{(w',S')}$ (because $R(B)_{\mathcal{A}}^{(w',S')}$ satisfies condition (R3) and either $h^{-1}[u] = \emptyset$ or $\{h^{-1}[u]\} \in S'$). One easily checks that $\mathsf{ev}{\circ}\langle f{\circ}h, a \rangle \upharpoonright h^{-1}[u] = \mathsf{ev}{\circ}\langle (f{\upharpoonright}u){\circ}h, a \rangle$ and, therefore, it follows that $\mathsf{ev} \circ \langle (f{\upharpoonright}u) \circ h, a \rangle \in R(B)_{\mathcal{A}}^{(w',S')}$ as required by condition (R3).

For verifying condition (R4) suppose that $f \in \mathcal{S}(A, B)^w$ and $P \in S$ with $f{\upharpoonright}u \in R([A{\to}B])_{\mathcal{A}}^{(w,S)}$ for all $u \in P$. We have to show that $f{\upharpoonright}\bigcup P \in R([A{\to}B])_{\mathcal{A}}^{(w,S)}$, too. For that purpose assume that $h : (w', S') \to (w, S)$ is a morphism in $\mathcal{A}$ and $a \in R(A)_{\mathcal{A}}^{(w',S')}$. We have to show that $\mathsf{ev}{\circ}\langle (f{\upharpoonright}\bigcup P) \circ h, a \rangle \in R(B)_{\mathcal{A}}^{(w',S')}$. As $\mathsf{ev} \circ \langle (f{\upharpoonright}\bigcup P) \circ h, a \rangle = \mathsf{ev} \circ \langle f \circ h, a \rangle \upharpoonright h^{-1}[\bigcup P]$ and $h^{-1}[\bigcup P] = \bigcup_{u \in P} h^{-1}[u]$ it suffices to show that $\mathsf{ev} \circ \langle f{\circ}h, a \rangle \upharpoonright h^{-1}[u] \in R(B)_{\mathcal{A}}^{(w',S')}$ for all $u \in P$ with $h^{-1}[u] \neq \emptyset$ (because $R(B)_{\mathcal{A}}^{(w',S')}$ satisfies condition (R4) and $\{h^{-1}[u] \mid u \in S\} \setminus \{\emptyset\} \in S'$). This, however, holds as $\{h^{-1}[u]\} \in S'$ for $u \in S$ with $h^{-1}[u] \neq \emptyset$ from which it follows by (R4) that $\mathsf{ev}{\circ}\langle f{\circ}h, a \rangle \upharpoonright h^{-1}[u] = \mathsf{ev}{\circ}\langle (f{\upharpoonright}u){\circ}h, a \rangle \in R(B)_{\mathcal{A}}^{(w',S')}$ since we have assumed that $f{\upharpoonright}u \in R([A{\to}B])_{\mathcal{A}}^{(w,S)}$.

It is immediate that the evaluation map $\mathsf{ev} : |[A{\to}B]| \times |A| \to |B| : (f, a) \mapsto f(a)$ preserves all relations and, therefore, is an $\mathcal{S}$-morphism. Suppose that $f : C \times A \to B$ is an $\mathcal{S}$-morphism. We show now that $\Lambda(f)$ with $\Lambda(f)(z)(x) = f(z, x)$ is an $\mathcal{S}$-morphism from C to $[A{\to}B]$. Suppose that $z \in |C|$. We have to show that $f(z, -) \in \mathcal{S}(A, B)$. Suppose $a \in R(A)_{\mathcal{A}}^{(w,S)}$. Then $\lambda i{\in}w.f(z, a(i)) = \lambda i{\in}w.f(\delta_w(z)(i), a(i)) = \mathsf{ev}{\circ}\langle \delta_w(z), a \rangle \in R(B)_{\mathcal{A}}^{(w,S)}$

because $\delta_w(z) \in R(C)_{\mathcal{A}}^{(w,S)}$ by (R2) and $a \in R(A)_{\mathcal{A}}^{(w,S)}$ by assumption. Thus, we have shown that $\Lambda(f)$ sends elements of $|C|$ to $\mathcal{S}(A,B)$. For showing that $\Lambda(f)$ preserves all relations suppose $c \in R(C)_{\mathcal{A}}^{(w,S)}$. We have to show that $\Lambda(f) \circ c \in R([A{\to}B])_{\mathcal{A}}^{(w,S)}$. Suppose that $h : (w', S') \to (w, S)$ is in $\mathcal{A}$ and $a \in R(A)_{\mathcal{A}}^{(w',S')}$. Then we have $\mathrm{ev} \circ \langle \Lambda(f) \circ c \circ h, a \rangle = f \circ \langle c \circ h, a \rangle \in R(B)_{\mathcal{A}}^{(w',S')}$ since $c \circ h \in R(C)_{\mathcal{A}}^{(w',S')}$ (as $R(C)_{\mathcal{A}}$ is a Kripke logical relation) and $a \in R(A)_{\mathcal{A}}^{(w',S')}$ by assumption. $\qquad\square$

Next we show that the usual interpretation of the arithmetic operations of PCF are actually $\mathcal{S}$-morphisms.

Lemma 11.8   *The usual interpretations of the arithmetic basic operations* zero, succ, pred *and* ifz *are morphisms in* $\mathcal{S}$ *when* **nat** *is interpreted by the object N as specified in Lemma 11.6.*

Proof.   This is trivial for zero, succ and pred and, therefore, left to the reader as an exercise(!). The case of cond $= [\![\text{ifz}]\!]$ is more subtle and illustrates the role of conditions (R3) and (R4) in Def. 11.3 and, therefore, we give it here in detail.

Suppose for that purpose that $f = \langle f_0, f_1, f_2 \rangle \in R(N \times N \times N)_{\mathcal{A}}^{(w,S)}$, i.e. that the $f_i \in R(N)_{\mathcal{A}}^{(w,S)}$. By definition of N (see Lemma 11.6) we know that $\{u, v\} \in S$ where $u := f_0^{-1}[\{0\}]$ and $v := f_0^{-1}[\mathbb{N}\backslash\{0\}]$. Thus, by condition (R3) of Definition 11.3 we have

$$(\text{cond} \circ f)\!\restriction\! u = f_1\!\restriction\! u \in R(N)_{\mathcal{A}}^{(w,S)} \quad \text{and} \quad (\text{cond} \circ f)\!\restriction\! v = f_2\!\restriction\! v \in R(N)_{\mathcal{A}}^{(w,S)}$$

from which it follows by condition (R4) of Definition 11.3 that

$$\text{cond} \circ f = (\text{cond} \circ f)\!\restriction\! u \cup v \in R(N)_{\mathcal{A}}^{(w,S)}$$

as desired. $\qquad\square$

Now we have assembled enough information to show that we get a model for PCF in $\mathcal{S}$ when interpreting **nat** as N.

Theorem 11.9   *The category $\mathcal{S}$ is a Λ-category when endowing the hom-sets $\mathcal{S}(A,B)$ with the pointwise ordering. Moreover, $\mathcal{S}$ is an order extensional model of* PCF *when interpreting* **nat** *as in Lemma 11.6 and the basic arithmetic operations as in the Scott model.*

Proof.   One easily checks that when defining pairing $\langle -, - \rangle$ and functional abstraction Λ as usual[1] this gives rise to a Λ-category (in the sense

---

[1]namely, as $\langle f, g \rangle(x) = \langle f(x), g(x) \rangle$ and $\Lambda(f)(x)(y) = f(x,y)$

of Definition 10.2). That $\mathcal{S}$ is order extensional follows from the pointwise definition of order on the hom-sets of $\mathcal{S}$ and the fact that for all sequential domains A all constant functions $a : \{\bot\} \to |A|$ are $\mathcal{S}$-morphisms (because by condition (R2) of Definition 11.3 each $R(B)_A^{(w,\mathcal{S})}$ contains all constant maps from w to $|B|$), i.e. we have enough global elements of each A available to distinguish different functions from A to B. From Lemma 11.6 we know that N is a sequemtial domain and from Lemma 11.8 we know that the usual interpreations of the basic arithmetic operations actually live in $\mathcal{S}$ as morphisms. Finally, we have least fixpoints as in any Λ-category (see Theorem 10.4). That least fixpoints are computed as usual by sending f to $\bigsqcup_{n \in \mathbb{N}} f^n(\bot)$ follows from Theorem 10.4 and the fact that in $\mathcal{S}$ pairing and application are constructed as usual. $\qquad\square$

In a sense the model $\mathcal{S}$ appears as sort of a "restriction of the Scott model to the sequential objects and their limits". This, however, is literally true only for first order types as for types of higher order one cannot really compare the functions because their domains of definition are too different. One may, however, define Kripke logical relations on the full Scott model and then verify (see [Liguoro 1996]) that the objects of the Scott model that are invariant under all these Kripke logical relations are precisely those elements of the Scott model which are the directed supremum of its approximating PCF definable compact elements, i.e. one may characterise by invariance properties the closure under directed suprema of the PCF definable elements of the Scott model. Instead, in the current chapter we have cut down "on the fly" all types to the closure of the PCF definable elements as we shall show in the following Chapter 12.

Chapter 12

The Model of PCF in $\mathcal{S}$ is Fully Abstract

From the main result of Chapter 10 we know that for order extensional models full abstraction is equivalent to PCF definability of all compact elements of all PCF types. For showing that all compact elements of interpretations of PCF types in the model $\mathcal{S}$ are actually definable in PCF we will examine the Kripke logical relations at certain varying arities $\mathcal{A}_n$ which will be defined subsequently after first fixing some notation in the following definition.

Definition 12.1 For every PCF type σ let A^σ be its interpretation in the PCF model in $\mathcal{S}$. For $n \in \mathbb{N}$ let $h_n^\sigma : A^\sigma \to A^\sigma$ be the interpretation of ψ_n^σ in $\mathcal{S}$. Like in the Scott model the h_n^σ form an ascending chain of finitary projections whose supremum is id_{A^σ}. We write A_n^σ for $h_n^\sigma[A^\sigma]$, the image of h_n^σ. If $\Gamma \equiv x_1{:}\sigma_1, \ldots, x_k{:}\sigma_k$ then we write A^Γ for $A^{\sigma_1} \times \ldots \times A^{\sigma_k}$, h_n^Γ for $h_n^{\sigma_1} \times \ldots \times h_n^{\sigma_k}$ and A_n^Γ for the image of h_n^Γ.

As there are only countably many contexts Γ and each $\mathcal{K}(A^\Gamma) = \bigcup_{n \in \mathbb{N}} A_n^\Gamma$ is countable we may choose an arbitrary, but fixed bijection between $\mathbb{N}$ and the (disjoint) union of all $\mathcal{K}(A^\Gamma)$. This bijection allows us to identify finite subsets of $\mathcal{K}(A^\Gamma)$ with (certain) finite subsets of $\mathbb{N}$. This identification will tacitly apply subsequently without further mention. $\Diamond$

We now define the sequence $\mathcal{A}_n$ of arities needed for the proof of full abstraction.

Definition 12.2 For a PCF context Γ and $n \in \mathbb{N}$ the ssp $\Gamma_n = (w_{\Gamma_n}, S_{\Gamma_n})$ is defined as follows

- w_{Γ_n} is A_n^Γ considered as a subset of $\mathbb{N}$
- $P \in S_{\Gamma_n}$ iff there is a term $\Gamma \vdash M : \iota$ such that

$$P = \{\, [\![\Gamma \vdash M]\!]^{-1}[\{i\}] \cap w_{\Gamma_n} \mid i \in \mathbb{N} \,\} \setminus \{\emptyset\} \,.$$

Now the arities $\mathcal{A}_n$ are defined as the subcategories of **SSP** whose objects are the Γ_n for arbitrary PCF contexts Γ and $n \in \mathbb{N}$ and whose morphisms are the projections

$$(\Gamma, \Delta)_n \to \Gamma_n : (\gamma, \delta) \mapsto \gamma$$

which, obviously, are closed under composition. $\Diamond$

Of course, one has to verify that the Γ_n are actually ssp's but this follows quite straightforwardly from closure properties of PCF terms of type ι in context Γ ensuring the requirements (SSP1)–(SSP4) of Definition 11.2 (see discussion after Def. 11.2). That the $\mathcal{A}_n$ are actually arities, i.e. that the projections $(\Gamma, \Delta)_n \to \Gamma_n$ are **SSP**-morphisms, follows from the fact that $\Gamma, \Delta \vdash M : \iota$ whenever $\Gamma \vdash M : \iota$.

Now we prove the main lemma ensuring full abstraction for the PCF model in $\mathcal{S}$ where **nat** is interpreted as the sequential domain N introduced in Lemma 11.6.

Lemma 12.3   *Let σ be a PCF type and $n \in \mathbb{N}$. Then for all contexts Γ and $f : A_n^\Gamma \to A_n^\sigma$ the following two conditions are equivalent*

(1) $f \in R(A^\sigma)_{\mathcal{A}_n}^{\Gamma_n}$
(2) $f = [\![\Gamma \vdash M]\!] \restriction A_n^\Gamma$ *for some term* $\Gamma \vdash M : \sigma$.

Proof.   The proof proceeds by induction on the structure of σ.

The base case is trivial because from the definition of S_{Γ_n} it is immediate that $R(A^\iota)_{\mathcal{A}_n}^{\Gamma_n} = R(N)_{\mathcal{A}_n}^{\Gamma_n}$ contains precisely the PCF-definable functions from A_n^Γ to $\mathbb{N}_\perp$.

Now suppose as induction hypotheses that conditions (1) and (2) are equivalent for the types σ and τ. For proving that (1) and (2) are equivalent for the type $\sigma \to \tau$ suppose $f : A_n^\Gamma \to A_n^{\sigma \to \tau}$.

For showing that (1) implies (2) suppose that $f \in R(A^{\sigma \to \tau})_{\mathcal{A}_n}^{\Gamma_n}$. Then we have

$$\mathrm{ev} \circ \langle f \circ \pi_1, \pi_2 \rangle \in R(A^\tau)_{\mathcal{A}_n}^{(\Gamma, x:\sigma)_n}$$

(where x is some fresh variable not declared in Γ) due to the definition of exponentials in $\mathcal{S}$ because π_1 is a morphism in $\mathcal{A}_n$ from $(\Gamma, x{:}\sigma)_n$ to Γ_n and $\pi_2 \in R(A^\sigma)_{\mathcal{A}_n}^{(\Gamma, x:\sigma)_n}$ by the induction hypothesis for σ since π_2 arises as interpretation of the term $\Gamma, x{:}\sigma \vdash x : \sigma$. Now from the induction hypothesis for τ it follows that

$$\mathrm{ev} \circ \langle f \circ \pi_1, \pi_2 \rangle = [\![\Gamma, x{:}\sigma \vdash M]\!] \restriction A_n^{\Gamma, x:\sigma}$$

for some PCF term $\Gamma, x{:}\sigma \vdash M : \tau$. But then we have

$$f = [\![\Gamma \vdash \psi_n^{\sigma \to \tau}(\lambda x{:}\sigma.M)]\!] \upharpoonright A_n^\Gamma$$

as desired.

For showing that (2) implies (1) assume that $f = [\![\Gamma \vdash M]\!] \upharpoonright A_n^\Gamma$ for some PCF term $\Gamma \vdash M : \sigma \to \tau$. For showing that $f \in R(A^{\sigma \to \tau})_{\mathcal{A}_n}^{\Gamma_n}$ consider an arbitrary morphism $\pi : (\Gamma, \Delta)_n \to \Gamma_n$ in $\mathcal{A}_n$ and an arbitrary element $a \in R(A^\sigma)_{\mathcal{A}_n}^{(\Gamma, \Delta)_n}$. By the induction hypothesis for σ there exists a PCF term $\Gamma, \Delta \vdash N : \sigma$ with $h_n^\sigma \circ a = [\![\Gamma, \Delta \vdash N]\!] \upharpoonright A_n^{\Gamma, \Delta}$ as $h_n^\sigma \circ a : A_n^{\Gamma, \Delta} \to A_n^\sigma$ is in $R(A^\sigma)_{\mathcal{A}_n}^{(\Gamma, \Delta)_n}$ because h_n^σ is PCF-definable and thus an $\mathcal{S}$-morphism. Now by induction hypothesis on τ we have

$$\begin{aligned}
\mathsf{ev} \circ \langle f \circ \pi, a \rangle &= \mathsf{ev} \circ \langle h_n^{\sigma \to \tau} \circ f \circ \pi, a \rangle = \\
&= \mathsf{ev} \circ \langle h_n^{\sigma \to \tau} \circ f \circ \pi, h_n^\sigma \circ a \rangle = \\
&= \mathsf{ev} \circ \langle f \circ \pi, h_n^\sigma \circ a \rangle = \\
&= [\![\Gamma, \Delta \vdash M(N)]\!] \upharpoonright A_n^{\Gamma, \Delta} \in R(A^\tau)_{\mathcal{A}_n}^{(\Gamma, \Delta)_n}
\end{aligned}$$

as desired. $\qquad\qquad\qquad\qquad\qquad\qquad\qquad\qquad\qquad\qquad\qquad\quad\square$

Now we can prove the desired full abstraction result for $\mathcal{S}$.

Theorem 12.4 (Full Abstraction for Sequential Domains)
For every PCF *type* σ *every compact element of* A^σ *is definable in* PCF. *Thus, the model of* PCF *in* $\mathcal{S}$ *with* $[\![\mathbf{nat}]\!] = N$ *is fully abstract.*

Proof. Suppose that $a \in A^\sigma$ is compact, i.e. $a \in A_n^\sigma$ for some $n \in \mathbb{N}$. Then the constant function $c_a : A_n^{\langle\rangle} \to A_n^\sigma$ with value a (where $\langle\rangle$ is the empty context) is an element of $R(A_n^\sigma)_{\mathcal{A}_n}^{\langle\rangle_n}$. Thus, by Lemma 12.3 there exists a PCF term $\langle\rangle \vdash M : \sigma$ with $c_a = [\![\langle\rangle \vdash M]\!] \upharpoonright \langle\rangle_n$, i.e. $a = [\![M]\!]$. Thus a is PCF definable as desired.

Now full abstraction of the PCF model in $\mathcal{S}$ follows from Theorem 10.11 since the model is order extensional and all compact elements of PCF types are PCF definable. $\qquad\qquad\qquad\qquad\qquad\qquad\qquad\qquad\qquad\qquad\quad\square$

The construction of the PCF model in $\mathcal{S}$ can be understood as a complete characterisation via *relational invariants* of PCF definability up to closure under directed suprema since we have shown that a Scott continuous function $f : A^\sigma \to A^\tau$ appears as supremum of a chain of PCF definable functions of type $\sigma \to \tau$ if and only if f preserves all Kripke logical relations. Thus, the relational invariants as given by the collection of all Kripke logical relations are sufficient for capturing PCF definability up to closure under directed suprema.

Notice, moreover, that the category S hosts also fully abstract models of functional programming languages with recursive types as has been investigated in detail in [Marz 2000].

It might be disappointing that we have used so many relational invariants for characterising PCF-definability up to closure under directed suprema. However, one cannot expect anything dramatically simpler because it has been shown by R. Loader [Loader 2001] that already for finitary PCF, i.e. PCF with booleans as its single base type instead of natural numbers, *observational equivalence is undecidable*. This forever refutes the original hope of characterising PCF-definability in terms of preservation of only finitely many relations.

Chapter 13

Computability in Domains

The intention of Domain Theory is to provide a mathematical semantics of computation. However, in the Scott model D for PCF (and also in the sequential domains model S) the type $D_{\iota\to\iota}$ does contain (strict) functions that are not computable. The aim of this chapter is to remedy this shortcoming by defining domains endowed with a notion of *computability*, the so-called *effectively given domains*.[1] As a byproduct we will arrive at a notion of *higher type computability* which is usually neglected[2] in the main stream of recursion-theoretic literature. In this chapter we will concentrate exclusively on Scott domains because for other notions of domains (as e.g. stable domains à la Berry and our sequential domains) there arise intrinsic difficulties when trying to extend the notion of computability to higher types. A more detailed account of Computability in Domains can be found in [Griffor et.al. 1994]. For background information about elementary recursion theory we refer the reader to the classic text by H. Rogers [Rogers 1987].

Before giving the precise definition of *effectively given domain* we recall a few basic facts about Scott domains, i.e. bounded complete countably algebraic domains. The most important property of Scott domains is that they are closed under exponentiation. Moreover, for Scott domains D and E the *basis* $\mathcal{K}([D\to E])$ of $[D\to E]$ can be described explicitly in terms of the bases $\mathcal{K}(D)$ and $\mathcal{K}(E)$ in the following way. First observe that a (countably) algebraic domain D is bounded complete if and only if every finite subset of $\mathcal{K}(D)$ has a supremum in D which necessarily is compact, too. Based on

---

[1]Notice, however, that effectively given domains need not be effectively isomorphic even if their underlying domains are isomorphic. For an explicit counterexample see [Kanda and Park 1979].

[2]Although some attention is sometimes paid to the second order case when discussing effective operations and effective operators as e.g. in [Rogers 1987].

this fact one easily shows that the compact elements of $[D \to E]$ are precisely those of the form

$$\bigsqcup_{i=1}^{n} [e_i, e_i']$$

where the e_i and e_i' are elements of $\mathcal{K}(D)$ and $\mathcal{K}(E)$, respectively, and the so-called step functions $[e_i, e_i']$ are defined as

$$[e_i, e_i'](x) = \begin{cases} e_i' & \text{if } e_i \sqsubseteq x \\ \bot & \text{otherwise} \end{cases}$$

provided the following *consistency* property holds: for every $I \subseteq \{1, \ldots, n\}$ with $\{e_i \mid i \in I\}$ bounded in D the set $\{e_i' \mid i \in I\}$ is bounded in E. Obviously, we have

$$\left(\bigsqcup_{i=1}^{n} [e_i, e_i']\right)(x) = \bigsqcup \{e_i' \mid e_i \sqsubseteq x\}$$

for $x \in D$. Based on these observations one easily checks that

$$[\tilde{e}, \tilde{e}'] \sqsubseteq \bigsqcup_{i=1}^{n} [e_i, e_i'] \quad \text{iff} \quad \tilde{e}' \sqsubseteq \bigsqcup \{e_i' \mid e_i \sqsubseteq \tilde{e}\} \ .$$

Accordingly, we have

$$\bigsqcup_{j=1}^{m} [\tilde{e}_j, \tilde{e}_j'] \sqsubseteq \bigsqcup_{i=1}^{n} [e_i, e_i'] \quad \text{iff} \quad \forall j \in \{1, \ldots, m\}. \ \tilde{e}_j' \sqsubseteq \bigsqcup \{e_i' \mid e_i \sqsubseteq \tilde{e}_j\} \ .$$

Now we are ready to define what is an effectively given (Scott) domain.

Definition 13.1   (Effectively Given Domain)
An *effectively given domain* is a pair (D, ε) such that D is a Scott domain and ε is an enumeration of $\mathcal{K}(D)$, i.e. $\varepsilon : \mathbb{N} \to \mathcal{K}(D)$ is a surjective function, satisfying the following two conditions

(1) $\uparrow\{\varepsilon(i) \mid i \in e_n\}$ is a decidable property of n (where e is some effective coding of $\mathcal{P}_{\mathsf{fin}}(\mathbb{N})$ and $\uparrow X$ stands for "X bounded from above")
(2) $\bigsqcup\{\varepsilon(i) \mid i \in e_n\} = \varepsilon(m)$ is a decidable relation between n and m.

An element $d \in D$ is called (D, ε)-*computable* or simply *computable* iff the set $\{n \in \mathbb{N} \mid \varepsilon(n) \sqsubseteq d\}$ is recursively enumerable (r.e.). We write $\mathsf{Comp}(D, \varepsilon)$ for the set of computable elements of (D, ε).   ◇

Next we define what are computable functions between effectively given domains.

Definition 13.2 (Computable Maps)
Let $(D_1, \varepsilon^{(1)})$ and $(D_2, \varepsilon^{(2)})$ be effectively given domains. A Scott continuous function $f : D_1 \to D_2$ is called *computable* (w.r.t. $\varepsilon^{(1)}$ and $\varepsilon^{(2)}$) iff the set[3] $\{\langle n, m \rangle \mid \varepsilon^{(2)}(m) \sqsubseteq f(\varepsilon^{(1)}(n))\}$ is r.e. To express that f is computable we write $f : (D_1, \varepsilon^{(1)}) \to (D_2, \varepsilon^{(2)})$. ◇

Notice that for an effectively given domain (D, ε) the relation $\varepsilon(n) \sqsubseteq \varepsilon(m)$ is decidable as it is equivalent to $\varepsilon(m) = \bigsqcup\{\varepsilon(n), \varepsilon(m)\}$. Accordingly, the relation $\varepsilon(n) = \varepsilon(m)$ is decidable, too.

From these observations the following two lemmas follow rather easily.

Lemma 13.3 *Computable maps between effectively given domains preserve computability of elements.*

Proof. Suppose $f : (D_1, \varepsilon^{(1)}) \to (D_2, \varepsilon^{(2)})$ and $x \in \mathsf{Comp}(D_1, \varepsilon^{(1)})$. Then the sets $A := \{n \in \mathbb{N} \mid \varepsilon^{(1)}(n) \sqsubseteq x\}$ and $B := \{\langle n, m \rangle \mid \varepsilon^{(2)}(m) \sqsubseteq f(\varepsilon^{(1)}(n))\}$ are r.e. Thus, the set $\{m \in \mathbb{N} \mid \exists n \in A.\ \langle n, m \rangle \in B\} = \{m \in \mathbb{N} \mid \varepsilon^{(2)}(m) \sqsubseteq f(x)\}$ is also r.e., i.e. $f(x)$ is computable. □

Lemma 13.4 *If (D, ε) is an effectively given domain then id_D is a computable map (w.r.t. ε). Moreover, computable maps are closed under composition, i.e. if $f : (D_1, \varepsilon^{(1)}) \to (D_2, \varepsilon^{(2)})$ and $g : (D_2, \varepsilon^{(2)}) \to (D_3, \varepsilon^{(3)})$ then $g \circ f : (D_1, \varepsilon^{(1)}) \to (D_3, \varepsilon^{(3)})$.*

Accordingly, effectively given domains and computable maps give rise to a category denoted as $\mathsf{Dom}_{\mathrm{eff}}$.

Proof. As by the remark after Def. 13.1 the relation $\varepsilon(n) \sqsubseteq \varepsilon(m)$ is decidable and thus also r.e. the map id_D is computable.

As f and g are computable the sets

$$A := \{\langle n, m \rangle \mid \varepsilon^{(2)}(m) \sqsubseteq f(\varepsilon^{(1)}(n))\} \quad B := \{\langle m, k \rangle \mid \varepsilon^{(3)}(k) \sqsubseteq g(\varepsilon^{(2)}(m))\}$$

are both r.e. By continuity of f and g we have $\varepsilon^{(3)}(k) \sqsubseteq (g \circ f)(\varepsilon^{(1)}(n))$ iff $\exists m \in \mathbb{N}.\ \varepsilon^{(2)}(m) \sqsubseteq f(\varepsilon^{(1)}(n)) \wedge \varepsilon^{(3)}(k) \sqsubseteq f(\varepsilon^{(2)}(m))$ iff $\exists m \in \mathbb{N}.\ \langle n, m \rangle \in A \wedge \langle m, k \rangle \in B$. As A and B are r.e. it follows that $\{\langle n, k \rangle \mid \varepsilon^{(3)}(k) \sqsubseteq (g \circ f)(\varepsilon^{(1)}(n))\}$ is r.e., too, and thus $g \circ f$ is computable.

As all identity maps on effectively given domains are computable and computable maps between effectively given domains are closed under composition they form a category $\mathsf{Dom}_{\mathrm{eff}}$. □

---

[3] usually called the (Scott) *graph of f*

The domain $\mathbb{N}_\perp$ can be made into an effectively given domain $N = (\mathbb{N}_\perp, \varepsilon^N)$ where $\varepsilon^N(0) = \perp$ and $\varepsilon^N(n{+}1) = n$. A terminal object 1 in $\mathsf{Dom}_{\mathsf{eff}}$ is given by $(\{\perp\}, \varepsilon^1)$ where $\varepsilon : \mathbb{N} \to \{\perp\}$ is the unique constant map with value $\perp$.

Lemma 13.5   *The category* $\mathsf{Dom}_{\mathsf{eff}}$ *is order extensional, i.e.* $f \sqsubseteq g$ *iff* $f \circ a \sqsubseteq g \circ a$ *for all* $a : 1 \to A$ *in* $\mathsf{Dom}_{\mathsf{eff}}$.

Proof.   Obvious from the fact that every compact element of an effectively given domain is in particular computable.   □

Next we show that the category of effectively given domains is cartesian closed.

Theorem 13.6   *The category* $\mathsf{Dom}_{\mathsf{eff}}$ *is cartesian closed.*

Proof.   A terminal object is given by $1 = (\{\perp\}, \varepsilon : \mathbb{N} \to \{\perp\})$.

Suppose $(D_1, \varepsilon^{(1)})$ and $(D_2, \varepsilon^{(2)})$ are effectively given domains. Their product is given by $([D_1 \times D_2], [\varepsilon^{(1)} \times \varepsilon^{(2)}])$ with $[\varepsilon^{(1)} \times \varepsilon^{(2)}](\langle n, m \rangle) = (\varepsilon^{(1)}(n), \varepsilon^{(2)}(m))$. Their exponential is given by $([D_1 \to D_2], [\varepsilon^{(1)} \to \varepsilon^{(2)}])$ for some appropriate numbering $[\varepsilon^{(1)} \to \varepsilon^{(2)}]$ which we describe next.

We call a finite subset A of $\mathbb{N}$ "consistent" iff the set $\{ [\varepsilon^{(1)}(i), \varepsilon^{(2)}(j)] \mid \langle i, j \rangle \in A \}$ of step functions is bounded in $[D_1 \to D_2]$. Due to the remarks before Def. 13.1 it is a decidable property of n whether e_n is consistent in this sense. Let $\widetilde{e}$ be an enumeration of the consistent finite subsets of $\mathbb{N}$ obtained from the canonical enumeration e of $\mathcal{P}_{\mathsf{fin}}(\mathbb{N})$ putting $\widetilde{e}_n = e_n$ if e_n is consistent and $\widetilde{e}_n = \emptyset$ otherwise. Now we define $[\varepsilon^{(1)} \to \varepsilon^{(2)}]$ as

$$[\varepsilon^{(1)} \to \varepsilon^{(2)}](n) = \{ [\varepsilon^{(1)}(i), \varepsilon^{(2)}(j)] \mid \langle i, j \rangle \in \widetilde{e}_n \}$$

for $n \in \mathbb{N}$. Next we argue why $[\varepsilon^{(1)} \to \varepsilon^{(2)}]$ satisfies requirements (1) and (2) of Def. 13.1. In order to decide whether $\uparrow \{ [\varepsilon^{(1)} \to \varepsilon^{(2)}](k) \mid k \in e_n \}$ decide whether $\bigcup_{k \in e_n} \widetilde{e}_k$ is consistent. Thus (1) holds for $[\varepsilon^{(1)} \to \varepsilon^{(2)}]$. In order to decide whether $\bigsqcup_{k \in e_n} [\varepsilon^{(1)} \to \varepsilon^{(2)}](k) = [\varepsilon^{(1)} \to \varepsilon^{(2)}](m)$ decide whether

$$\bigsqcup \{ [\varepsilon^{(1)}(i), \varepsilon^{(2)}(j)] \mid \langle i, j \rangle \in \widetilde{e}(k), k \in e_n \} = \bigsqcup \{ [\varepsilon^{(1)}(i), \varepsilon^{(2)}(j)] \mid \langle i, j \rangle \in \widetilde{e}_m \}$$

which can be done effectively due to the remarks before Def. 13.1. Thus (2) holds for $[\varepsilon^{(1)} \to \varepsilon^{(2)}]$.

One easily checks (exercise!) that the set

$$\{ \langle \langle n, m \rangle, k \rangle \mid \varepsilon^{(2)}(k) \sqsubseteq [\varepsilon^{(1)} \to \varepsilon^{(2)}](n)(\varepsilon^{(1)}(m)) \}$$

is r.e. from which it follows that the map $\mathsf{ev} : [D_1 {\to} D_2] \times D_1 \to D_2$ is computable.

Suppose $(D_3, \varepsilon^{(3)})$ is an effectively given domain and $f : D_3 \times D_1 \to D_2$ is computable, i.e. $F := \{\langle\langle n, m\rangle, k\rangle \mid \varepsilon^{(2)}(k) \sqsubseteq f(\varepsilon^{(3)}(n), \varepsilon^{(1)}(m))\}$ is r.e. Then the function $\Lambda(f) : D_3 \to [D_1 {\to} D_2] : z \mapsto (x \mapsto f(z, x))$ is computable because $[\varepsilon^{(1)} {\to} \varepsilon^{(2)}](m) \sqsubseteq \Lambda(f)(\varepsilon^{(3)}(n))$ iff $\forall \langle i, j\rangle \in \widetilde{e}_m .\ \langle\langle n, i\rangle, j\rangle \in F$ and the latter condition is obviously semidecidable since F is semidecidable by assumption.

This concludes the proof that $\big([D_1 {\to} D_2], [\varepsilon^{(1)} {\to} \varepsilon^{(2)}]\big)$ is actually the desired exponential in $\mathsf{Dom}_{\mathsf{eff}}$. $\qquad\square$

Next we show that taking least fixpoints is a computable operation.

Theorem 13.7 *For an effectively given domain (D, ε) the least fixpoint operator $\mu_D : [D {\to} D] \to D : f \mapsto \bigsqcup_{n \in \mathbb{N}} f^n(\bot)$ is computable.*

Proof. Recall from the proof of the previous Theorem 13.6 that a compact element $[\varepsilon {\to} \varepsilon](n) \in \mathcal{K}([D {\to} D])$ equals $\bigsqcup\{[\varepsilon(i), \varepsilon(j)] \mid \langle i, j\rangle \in \widetilde{e}(n)\}$. First observe that the set $\{k \in \mathbb{N} \mid \varepsilon(k) \sqsubseteq \mu_D([\varepsilon {\to} \varepsilon](n))\}$ is the least set I_n s.t.

(1) if $\varepsilon(k) = \bot_D$ then $k \in I_n$ and
(2) if $k \in I_n$, $\varepsilon(i) \sqsubseteq \varepsilon(k)$ and $\langle i, j\rangle \in \widetilde{e}_n$ then $j \in I_n$.

As I_n is r.e. *uniformly* in n the set

$$\{\langle n, m\rangle \mid m \in I_n\} = \{\langle n, m\rangle \mid \varepsilon(m) \sqsubseteq \mu_D([\varepsilon {\to} \varepsilon](n))\}$$

is r.e. and, accordingly, the map μ_D is computable. $\qquad\square$

Interpreting base type ι as $N = (\mathbb{N}_\bot, \varepsilon^N)$ gives rise to the observation that the interpretation of PCF in Scott domains factors through $\mathsf{Dom}_{\mathsf{eff}}$ because the basic operations zero, succ, pred and ifz on $\mathbb{N}_\bot$ are all computable and the interpretation of typed λ-calculus and recursion does not lead outside the effective world as guaranteed by Theorems 13.6 and 13.7, respectively. This model of PCF inside $\mathsf{Dom}_{\mathsf{eff}}$ is called the *effective Scott model*.

Lemma 13.8 *The interpretations of por and $\exists$ are computable elements of $N {\to} N {\to} N$ and $(N {\to} N) {\to} N$, respectively.*

Proof. One simply checks that in both cases the set of codes of approximating compact elements is r.e. (exercise!). $\qquad\square$

Thus, all terms of $\mathrm{PCF}^{++} \equiv \mathrm{PCF} + \mathsf{por} + \exists$ denote computable elements of the Scott model. We write $(D_\sigma, \varepsilon^\sigma)$ for the interpretation of PCF type σ in $\mathsf{Dom_{eff}}$ where base type ι gets interpreted as $N = (\mathbb{N}_\bot, \varepsilon^N)$.

In [Plotkin 1977] G. Plotkin has proved the following two remarkable theorems

Theorem 13.9 (Full Abstraction for PCF^+)
Every compact element of the Scott model of PCF *arises as interpretation of some closed term in* $\mathrm{PCF}^+ \equiv \mathrm{PCF} + \mathsf{por}$. *Thus, the Scott model is fully abstract for* PCF^+.

Theorem 13.10 (Universality for PCF^{++})
All computable elements of PCF *types arise as denotations of* PCF^{++} *terms. Thus, the effective Scott model is universal for* PCF^{++}.
Moreover, every object of the Scott model of PCF *is definable in* PCF^{++}_Ω, *i.e.* PCF^{++} *extended by constants (so-called "oracles") for all strict total functions of type* $\iota \to \iota$.

telling us that the effective Scott model is the best possible one for PCF^{++} and that it is precisely (the absence of) por which is responsible for the Scott model's lack of full abstraction for PCF. Moreover, Theorem 13.10 tells us that

PCF^{++} provides a *purely extensional* account of classical[4] recursion theory extended to higher types.

However, it will take several steps to prove these two classical theorems of Plotkin. First recall that in chapter 7 we invited the reader to prove that

Theorem 13.11 *In the Scott model for all* PCF *types* σ *the domain* D_σ *is coherently complete.*

Proof. Obviously, $\mathbb{N}_\bot$ is coherently complete and coherently complete cpo's are are closed under $\times$. Suppose D and E are domains and E is coherently complete. Then $[D \to E]$ is coherently complete, too, as if $F \subseteq [D \to E]$ is coherent then for all $x \in D$ the set $\{f(x) \mid f \in F\}$ is coherent, too, and thus has a supremum (as E is coherently complete by assumption). One easily checks (as in the proof of Lemma 3.5) that the function g with

$$g(x) = \bigsqcup \{f(x) \mid f \in F\} \qquad \text{for } x \in D$$

is continuous and thus the supremum of F w.r.t. the pointwise order. $\quad\square$

---

[4]as in classical recursion theory por and $\exists$ are available via *dove tailing*

If x and y are elements of a domain D we write $x{\uparrow}y$ for ${\uparrow}\{x,y\}$ and $x\#y$ for the negation of $x \uparrow y$.

Lemma 13.12   *Let D be a coherently complete Scott domain. Then for every $e \in \mathcal{K}(D)$ the function* $\mathrm{below}_e : D \to \mathbb{N}_\perp$ *with*

$$\mathrm{below}_e(d) = \begin{cases} 0 & \text{if } e \sqsubseteq d \\ 1 & \text{if } e\#d \\ \perp & \text{otherwise} \end{cases}$$

is continuous.

Proof.   Obviously, the function below_e is monotonic. Suppose $X \subseteq D$ is directed and $\mathrm{below}_e(x) = \perp$ for all $x \in X$. Then $e{\uparrow}x$ for all $x \in X$ from which it follows that $e \uparrow \bigsqcup X$ because $\{e\} \cup X$ is coherent and, thus, has a supremum. Moreover, we have $e \not\sqsubseteq \bigsqcup X$ as otherwise $e \sqsubseteq x$ for some $x \in X$ since e is assumed as compact. Thus $\mathrm{below}_e(\bigsqcup X) = \perp$ as well.   □

Lemma 13.13   (parallel conditional in PCF^+)
In PCF^+ *one can exhibit a term* $x{:}\iota, y{:}\iota, z{:}\iota \vdash$ **pif** x **then** y **else** z **fip** $: \iota$
whose interpretation is the ternary function pcond *on* $\mathbb{N}_\perp$ *defined as*

$$\mathrm{pcond}(x,y,z) = \begin{cases} y & \text{if } x = 0 \text{ or } y = z \in \mathbb{N} \\ z & \text{if } x = 1 \\ \perp & \text{otherwise} \end{cases}$$

for all $x,y,z \in \mathbb{N}_\perp$.

Proof.   First observe that in PCF^+ one can implement a function pand of type $\iota{\to}\iota{\to}\iota$ such that

$$\mathsf{pand}\, x\, y = \begin{cases} 0 & \text{if } x = 0 = y \\ 1 & \text{if } x = 1 \text{ or } y = 1 \\ \perp & \text{otherwise} \end{cases}$$

For the sake of readability we write $x \wedge y$ as an abbreviation for $\mathsf{pand}\, x\, y$ and $x \vee y$ as an abbreviation for $\mathsf{por}\, x\, y$. Using fixpoint operators we can define in PCF^+ a function $h : (\iota{\to}\iota) \to \iota \to \iota$ such that

$$h\, f\, x = \mathsf{ifz}(f(x), x, h(f)(\mathsf{succ}(x)))$$

for all $h \in D_{\iota\to\iota}$ and $x \in D_\iota$. Then the function $\mathrm{search} : (\iota{\to}\iota) \to \iota$ defined as

$$\mathrm{search}(f) = h\, f\, \mathsf{zero}$$

is also PCF^+ definable and implements unbounded search in $\mathbb{N}$, i.e., $\mathrm{search}(f)$ terminates iff there exists an $n \in \mathbb{N}$ such that $f(n) = 0$ and $f(k) \in \mathbb{N}\backslash\{0\}$ for all $k < n$ in which case $\mathrm{search}(f) = n$.

Now when defining **pif** x **then** y **else** z **fip** as

$$\mathrm{search}(\lambda w{:}\iota.\ (w{=}y \wedge w{=}z) \vee (w{=}y \wedge x{=}0) \vee (w{=}z \wedge x{=}1))$$

one easily checks that it satisfies its specification. $\square$

Now we can prove that the Scott model is fully abstract for PCF^+.

Proof (of Theorem 13.9) :
By induction on the structure of PCF types σ we show that the following three claims hold for all $e, e' \in \mathcal{K}(D_\sigma)$

(1) e is definable in PCF^+
(2) $[e, 0]$ is definable in PCF^+
(3) $[e, 0] \sqcup [e', 1]$ is definable in PCF^+ whenever $e \# e'$.

The claims obviously hold for base type ι. Suppose $\sigma = \sigma_1 {\to} \dots {\to} \sigma_k {\to} \iota$ and as induction hypothesis that (1)–(3) hold for the σ_i.

First some notation. For compact elements $e_i \in \mathcal{K}(D_{\sigma_i})$ and $n \in \mathbb{N}$ we write $[e_1, \dots, e_k, n]$ for the function $f \in D_\sigma$ with

$$f f_1 \dots f_k = \begin{cases} n & \text{if } e_i \sqsubseteq f_i \text{ for all } i \in \{1, \dots, k\} \\ \bot & \text{otherwise.} \end{cases}$$

Such functions are called *step functions*. Every $e \in \mathcal{K}(D_\sigma)$ is the supremum of a finite set of such step functions. One easily checks (exercise!) that there exists a least such set which we denote as F_e.

We proceed by induction on the size of $|F_e|$ and $|F_{e'}|$.

ad (1): If F_e is empty then e is denoted by Ω_σ. Thus, w.l.o.g. suppose $F_e \neq \emptyset$.

If there are $[e_1, \dots, e_k, n], [e'_1, \dots, e'_k, n'] \in F_e$ with $e_i \# e'_i$ for some i with $1 \leq i \leq k$ then by induction hypothesis there exists a PCF^+ term M denoting $[e_i, 0] \sqcup [e'_i, 1]$. By induction hypothesis $\bigsqcup F_e \backslash \{[e_1, \dots, e_k, n]\}$ and $\bigsqcup F_e \backslash \{[e'_1, \dots, e'_k, n']\}$ are definable by terms N_1 and N_2, respectively. But then $e = \bigsqcup F_e$ is definable by the PCF^+ term

$$\lambda f_1{:}\sigma_1.\dots.\lambda f_k{:}\sigma_k.\, \textbf{pif}\ M f_i\ \textbf{then}\ N_2 f_1 \dots f_k\ \textbf{else}\ N_1 f_1 \dots f_k\ \textbf{fip}$$

Otherwise for all $[e_1, \dots, e_k, n], [e'_1, \dots, e'_k, n'] \in F_e$ we have $n = n'$. Take $[e_1, \dots, e_k, n] \in F_e$. By induction hypothesis there are PCF^+

terms M_i denoting $[e_i, 0]$ for $i = 1, \ldots, k$ and a PCF^+ term N denoting $\bigsqcup F_e \setminus \{[e_1, \ldots, e_k, n]\}$. Then $e = \bigsqcup F_e$ is denoted by the PCF^+ term

$$\lambda f_1{:}\sigma_1 \ldots \lambda f_k{:}\sigma_k . \, \mathbf{pif} \, M_1 f_1 \wedge \cdots \wedge M_k f_k \, \mathbf{then} \, n \, \mathbf{else} \, N f_1 \ldots f_k \, \mathbf{fip}$$

where $\wedge$ stands for pand (as introduced in the proof of Lemma 13.13).

ad (2): If $F_e = \emptyset$ then $\lambda x{:}\sigma.0$ denotes $[e, 0]$. Otherwise take some $[e_1, \ldots, e_k, n] \in F$. By induction hypothesis there exist PCF^+ terms $M_1, \ldots, M_k, E$ and N denoting the compact elements $e_1, \ldots, e_k$, $[n, 0]$ and $\left[\bigsqcup F_e \setminus \{[e_1, \ldots, e_k, n]\}, 0\right]$, respectively. Then the PCF^+ term

$$\lambda f{:}\sigma. \, \mathsf{ifz}(E(f M_1 \ldots M_k), N f, \Omega_\iota)$$

denotes $[e, 0]$.

ad (3): Suppose $e \# e'$. Then there exist $[e_1, \ldots, e_k, n] \in F_e$ and $[e_1', \ldots, e_k', n'] \in F_{e'}$ such that $n \neq n'$ and $e_i \uparrow e_i'$ for $i = 1, \ldots, k$. By induction hypothesis there are PCF^+ terms $M_1, \ldots, M_k$ denoting $e_1 \sqcup e_1', \ldots, e_k \sqcup e_k'$, respectively. As (2) has already been established for σ there exist PCF^+ terms N and N' denoting $[e, 0]$ and $[e', 0]$, respectively. Let K be a PCF term denoting $[n, 0] \sqcup [n', 1]$. Then $[e, 0] \sqcup [e', 1]$ is denoted by

$$\lambda f{:}\sigma. \, \mathsf{ifz}(K(f M_1 \ldots M_k), N f, \mathsf{succ}(N' f))$$

which, obviously, is a PCF^+ term. □

For proving universality of PCF^{++} we have to observe that sort of a "continuous universal quantifier" is definable in PCF^{++}.

Lemma 13.14 *The function* $\forall : [\mathbb{N}_\perp \to \mathbb{N}_\perp] \to \mathbb{N}_\perp$ *with*

$$\forall(f) = \begin{cases} 0 & \text{if } f(\perp) = 0 \\ 1 & \text{if } f(n) = 1 \text{ for some } n \in \mathbb{N} \\ \perp & \text{otherwise} \end{cases}$$

is definable in PCF^{++}.

Proof. Let swap be a PCF term of type $\iota \to \iota$ with

$$\mathsf{swap}(x) = \begin{cases} 0 & \text{if } x = 1 \\ 1 & \text{if } x = 0 \\ x & \text{otherwise.} \end{cases}$$

Then the PCF^{++} term $\lambda f{:}\iota \to \iota. \, \mathsf{swap}(\exists(\lambda x{:}\iota. \, \mathsf{swap}(f(x))))$ denotes $\forall$. □

Moreover, we will employ the following extension of ifz and the parallel conditional to arbitrary types $\sigma = \sigma_1 \to \ldots \to \sigma_k \to \iota$: if M is of type ι and N_1 and N_2 are of type σ (relative to some typing context) then we write $\mathsf{ifz}(M, N_1, N_2)$ as a shorthand for

$$\lambda x_1{:}\sigma_1. \ldots . \lambda x_n{:}\sigma_n.\mathsf{ifz}(M, N_1 x_1 \ldots x_n, M_2 x_1 \ldots x_n)$$

and **pif** M **then** N_1 **else** N_2 **fip** as a shorthand for

$$\lambda x_1{:}\sigma_1. \ldots . \lambda x_n{:}\sigma_n.\textbf{pif } M \textbf{ then } N_1 x_1 \ldots x_n \textbf{ else } M_2 x_1 \ldots x_n \textbf{ fip} \ .$$

We leave it as an exercise(!) to check that

$$\textbf{pif } \Omega_\iota \textbf{ then } N_1 \textbf{ else } N_2 \textbf{ fip} = N_1 \sqcap N_2$$

which fact we will tacitly use in the subsequent proof of Theorem 13.10 which is due to [Escardó 1997] and easier to follow than the original proof in [Plotkin 1977].

Proof (of Theorem 13.10) :
By induction on the structure of PCF types we will show that for every type σ there exist PCF^{++} terms $\mathrm{join}^\sigma : \iota \to \sigma \to \sigma$, $\mathrm{up}^\sigma : \iota \to \iota \to \sigma$ and $\mathrm{below}^\sigma : \iota \to \sigma \to \iota$ whose denotations (also denoted as join^σ, up^σ and below^σ) satisfy the requirements that

(1) for all $d \in D_\sigma$ and $n \in \mathbb{N}$

   (i) $\mathrm{join}^\sigma(\bot)(d) = \bot$ and
   (ii) $\mathrm{join}^\sigma(n)(d) = \varepsilon_n^\sigma \sqcup d$ whenever $\varepsilon_n^\sigma \uparrow d$

(2) for every $n \in \mathbb{N}$ the set $\{d \in \mathsf{Comp}(D_\sigma, \varepsilon^\sigma) \mid \varepsilon_n^\sigma \sqsubseteq d\}$ is enumerated by the function $\mathrm{up}^\sigma(n) : \mathbb{N}_\bot \to D_\sigma \mathrm{d}$

(3) for all $n \in \mathbb{N}$ and $d \in D_\sigma$

$$\mathrm{below}^\sigma(n)(d) = \begin{cases} 0 & \text{if } \varepsilon_n^\sigma \sqsubseteq d \\ 1 & \text{if } \varepsilon_n^\sigma \mathbin{\#} d \\ \bot & \text{otherwise} \end{cases}$$

i.e. $\mathrm{below}^\sigma(n)$ implements the function $\mathrm{below}_{\varepsilon_n^\sigma}$ of Lemma 13.12.

Before verifying the existence of such terms we show how they allow one to prove that all computable elements of D_σ can be denoted by PCF^{++} terms.

Suppose $d \in \mathsf{Comp}(D_\sigma, \varepsilon^\sigma)$. Then there exists a PCF term $g_d : \iota \to \iota$ with $\{g_d(n) \mid n \in \mathbb{N}\} = \{n \in \mathbb{N} \mid \varepsilon_n^\sigma \sqsubseteq d\}$. We write h_d for the PCF^{++} term $\lambda k{:}\iota.\mathsf{join}^\sigma(g_d(k))$ of type $\iota \to \sigma \to \sigma$. Let Φ_σ be the PCF term

$$\lambda F{:}(\iota \to \sigma \to \sigma) \to \sigma.\lambda f{:}\iota \to \sigma \to \sigma.f(0)(F(\lambda n{:}\iota.f(\mathsf{succ}(n))))$$

of type $((\iota \to \sigma \to \sigma) \to \sigma) \to (\iota \to \sigma \to \sigma) \to \sigma$. Then we have

$$
\begin{aligned}
\mathsf{Y}(\Phi_\sigma)(h_d) &= \bigsqcup\nolimits_{n \in \mathbb{N}} \Phi_\sigma^n(\bot)(h_d) = \\
&= \bigsqcup\nolimits_{n \in \mathbb{N}} h_d(0) \circ \cdots \circ h_d(n)(\bot) = \\
&= \bigsqcup\nolimits_{n \in \mathbb{N}} \varepsilon_{g_d(0)}^\sigma \sqcup \cdots \sqcup \varepsilon_{g_d(n)}^\sigma = \\
&= d
\end{aligned}
$$

where the last equality holds as g_d enumerates the codes of finite approximations to d. Thus, the PCF^{++} term $\mathsf{Y}(\Phi_\sigma)(h_d)$ denotes d and hence d is definable in PCF^{++}.

For arbitrary $d \in D_\sigma$ there still exists a total function $g_d : \mathbb{N} \to \mathbb{N}$ enumerating $\{n \in \mathbb{N} \mid \varepsilon_n^\sigma \sqsubseteq d\}$ which, however, in general will not be recursive anymore. Nevertheless, we still have $d = \mathsf{Y}(\Phi)(\lambda k{:}\iota.\mathsf{join}^\sigma(g_d(k)))$ and hence d is definable in PCF$_\Omega^{++}$ (using the oracle g_d).

Now we will turn back to the task of exhibiting terms join$^\sigma$, up$^\sigma$ and below$^\sigma$ satisfying requirements (1)–(3). We proceed by induction on the structure of σ. For base type ι the claim is obvious. Suppose as induction hypotheses that the claim holds already for σ and τ.

There is a PCF^{++} definable function sjoin : $\mathbb{N}_\bot \to D_{\sigma \to \tau} \to D_{\sigma \to \tau}$ with

$$\mathsf{sjoin}(\langle n, m \rangle)(f)(d) = \mathbf{pif}\ \mathsf{below}^\sigma(n)(d)\ \mathbf{then}\ \mathsf{join}^\tau(m)(f(d))\ \mathbf{else}\ f(d)\ \mathbf{fip}$$

for all $n, m \in \mathbb{N}$. If $[\varepsilon_n^\sigma, \varepsilon_m^\tau] \uparrow f$ then $\mathsf{sjoin}(\langle n, m \rangle)(f) = [\varepsilon_n^\sigma, \varepsilon_m^\tau] \sqcup f$ as can be seen from the following case analysis

i) if $\varepsilon_n^\sigma \sqsubseteq d$ then $\mathsf{sjoin}(\langle n, m \rangle)(f)(d) = \varepsilon_m^\tau \sqcup f(d) = ([\varepsilon_n^\sigma, \varepsilon_m^\tau] \sqcup f)(d)$
ii) if $\varepsilon_n^\sigma \uparrow d$ but not $\varepsilon_n^\sigma \sqsubseteq d$ then $\varepsilon_m^\tau \uparrow f(d)$ and below$^\sigma(n)(d) = \bot$ and, thus, we have $\mathsf{sjoin}(\langle n, m \rangle)(f)(d) = \mathsf{join}^\tau(m)(f(d)) \sqcap f(d) = f(d) = ([\varepsilon_n^\sigma, \varepsilon_m^\tau] \sqcup f)(d)$ because $f(d) \sqsubseteq \varepsilon_m^\tau \sqcup f(d) = \mathsf{join}^\tau(m)(f(d))$ as guaranteed by $\varepsilon_m^\tau \uparrow f(d)$ and the induction hypothesis for join$^\tau$
iii) if $\varepsilon_n^\sigma \# d$ then $\mathsf{sjoin}(\langle n, m \rangle)(f)(d) = f(d) = ([\varepsilon_n^\sigma, \varepsilon_m^\tau] \sqcup f)(d)$.

Now join$^{\sigma \to \tau}$ is PCF definable from sjoin putting

$$\mathsf{join}^{\sigma \to \tau}(n)(f)(d) = \mathsf{ifz}(b^{\sigma \to \tau}(n), f(d), \mathsf{sjoin}(\delta(n))(\mathsf{join}^{\sigma \to \tau}(\rho(n))(f))(d))$$

where $b^{\sigma\to\tau}$ is a recursive function deciding $\varepsilon_n^{\sigma\to\tau} = \bot$ and δ and ρ are recursive functions such that for $n \neq 0$, $\delta(n)$ is the least element of e_n and $e_{\rho(n)} = e_n \backslash \{\delta(n)\}$.

For defining up^σ first consider the PCF^{++} term

$$\mathrm{enum}_1^\sigma \equiv \lambda f{:}\iota{\to}\iota.\mathsf{Y}(\Phi_\sigma)(\lambda k{:}\iota.\mathrm{join}^\sigma(f(k)))$$

where Φ_σ is defined as above. Obviously, if $d \in D_\sigma$ and $f \in D_{\iota\to\iota}$ with $\{f(n) \mid n \in \mathbb{N}\} = \{n \in \mathbb{N} \mid \varepsilon_n^\sigma \sqsubseteq d\}$ then $d = \mathrm{enum}_1^\sigma(f)$. Let U be a PCF term of type $\iota{\to}\iota{\to}\iota$ such that $\{U(n) \mid n \in \mathbb{N}\}$ is the set of all unary partial recursive functions.[5] Then $\lambda x{:}\iota.\mathrm{enum}_1^\sigma(U(x))$ enumerates $\mathsf{Comp}(D_\sigma, \varepsilon^\sigma)$. Thus, the PCF^{++} definable function

$$\mathrm{up}^{\sigma\to\tau}(n)(m) = \mathrm{join}^{\sigma\to\tau}(n)(\mathrm{enum}_1^{\sigma\to\tau}(m))$$

satisfies the requirement formulated in (2).

Finally $\mathrm{below}^{\sigma\to\tau}$ can be defined as

$$\mathrm{below}^{\sigma\to\tau}(n)(f) = \forall_{\langle i,j\rangle\in e_n}\forall(\lambda k{:}\iota.\,\mathrm{below}^\tau(j)(f(\mathrm{up}^\sigma(i)(k))))$$

where $\forall$ is the continuous universal quantifier of Lemma 13.14 and the quantification over e_n can be expressed in terms of pand (as introduced in the proof of Lemma 13.13). As below^τ and up^σ are PCF^{++} definable by induction hypothesis the function $\mathrm{below}^{\sigma\to\tau}$ is PCF^{++} definable as well. $\square$

See pp. 250-251 of [Plotkin 1977] for a syntactic argument showing that $\exists$ is not PCF$^+$ definable. That already PCF$^+$ suffices for denoting all compact elements of the Scott model we consider rather as a mere curiosity. But the following two facts one certainly should remember.

- The Scott model is not fully abstract for PCF as the latter lacks parallel features.
- If one adds por and $\exists$ to PCF, i.e. extensional parallel features providing all the benefits of *dove tailing*, then one can denote all computable elements of the Scott model, i.e. the language PCF^{++} is universal for the effective Scott model $\mathsf{Dom}_{\mathrm{eff}}$.

---

[5]where a unary partial function f on $\mathbb{N}$ is identified with the element $\tilde{f} \in D_{\iota\to\iota}$ defined as

$$\tilde{f}(x) = \begin{cases} f(x) & \text{if } x \in \mathbb{N} \text{ and } f(x){\downarrow} \\ \bot & \text{otherwise} \end{cases}$$

The above universality theorem has been extended to PCF^{++} with recursive types in [Streicher 1994] by showing that every recursive type appears as a PCF^{++} definable retract of the type $\iota \to \iota$.[6] However, it is an open problem whether in the category S of sequential domains there is a PCF type υ such that every (recursive) type appears as PCF definable retract of υ.

As we have seen there is a smooth notion of computability for Scott domains. When trying to achieve something similar for the sequential domains (introduced in Chapter 11) one runs into the following problems.

Consider the projections $\pi_1, \pi_2 : \Sigma \times \Sigma \to \Sigma$ (where Σ is the 2-element lattice $\{\bot, \top\}$). Then in any sequential model the pointwise supremum $\pi_1 \sqcup \pi_2$ does not exist as it is the inherently parallel supremum operation $\vee : \Sigma \times \Sigma \to \Sigma$. The only (pointwise) upper bound of π_1 and π_2 in a sequential model is the constant map with value $\top$. Thus, in order extensional sequential models suprema in function spaces are not pointwise. Moreover, it follows from [Loader 2001] that already in the fully abstract model for finitary PCF, i.e. over base type **bool**, where all elements are compact, it is not decidable whether an element is the supremum of a finite set of elements.

Any attempt to develop a notion of computability for sequential domains is hampered by the following observation. Let K be a non-recursive, but r.e. set of natural numbers (e.g. $K = \{n \in \mathbb{N} \mid \{n\}(n)\downarrow\}$). For every $n \in \mathbb{N}$ consider the function $f_n : \Sigma \to [\mathbb{N} \to \Sigma]$ defined as

$$f_n(u) = \begin{cases} \mathbb{N} & \text{if } u = \top \\ \{k \in K \mid k < n\} & \text{otherwise.} \end{cases}$$

Obviously, each f_n is sequential and computable as it can be implemented e.g. by an obvious ML program of type unit$->$nat$->$unit. The sequence $(f_n)_{n \in \mathbb{N}}$ is effective and ascending w.r.t. the extensional order but, nevertheless, its limit $f = \bigsqcup f_n$ as given by

$$f(u) = \begin{cases} \mathbb{N} & \text{if } u = \top \\ K & \text{otherwise} \end{cases}$$

---

[6]Thus, in case of the effective Scott model, i.e. higher type computability *in the sense of recursion theory*, higher types are just a "figure of speech" because they can all be simulated within $\iota \to \iota$, i.e. the partial recursive functions (together with all constant functions). Nevertheless, higher types are very convenient from a "stylistic" point of view.

is *not sequentially computable* since

$$f(u)(n) = \top \quad \text{iff} \quad n \in K \vee u{=}\top$$

and the right hand side obviously requires parallel evaluation. Thus, for sequential domains the computable elements are not closed under suprema of effective chains of effective elements.

However, via *realizability* one may construct universal models for sequential languages (even with recursive types), see [Rohr 2002].[7]

But for the stable model of PCF (see [Amadio and Curien 1998]) a notion computability has been successfully developed by A. Asperti in [Asperti 1990]. This is possible because the sequence $(f_n)_{n \in \mathbb{N}}$ considered above is *not increasing w.r.t. the stable ordering*.

We conclude this chapter with some more recursion theoretic considerations.

Principal Numberings and the Myhill-Shepherdson Theorem

For an effectively given domain we will define a notion of *principal numbering* of $\mathsf{Comp}(D, \varepsilon)$ which in a certain sense will be optimal. For this purpose recall the Gödel numbering W of r.e. sets of natural numbers where

$$W_e = \{n \in \mathbb{N} \mid \{e\}(n){\downarrow}\}$$

see e.g. [Rogers 1987].

Definition 13.15   (principal numbering)
Let (D, ε) be an effectively given domain. A *principal numbering* of (D, ε) is a surjective function $\zeta : \mathbb{N} \to \mathsf{Comp}(D, \varepsilon)$ such that there exist total recursive functions f and g satisfying the conditions

$(\zeta 1) \quad W_{f(n)} = \{k \in \mathbb{N} \mid \varepsilon_k \sqsubseteq \zeta(n)\}$
$(\zeta 2) \quad \zeta(g(n)) = \bigsqcup \varepsilon[W_n]$ whenever $\varepsilon[W_n]$ is directed.

---

[7]See also John Longley's treatise [Longley 2002] on *Sequentially Realizable Functionals* where he constructs a universal model for PCF+H where H is a non-order-extensional but sequential constant of type 2 having a somewhat complicated operational semantics. Longley's sequentially realizable functionals are equivalent to the extensional collapse of Curien's *Sequential Algorithms* (see [Amadio and Curien 1998]) providing a non-extensional model for PCF.

But notice that this notion of sequentiality is more liberal than the one arising from PCF definability as studied in this book.

A *computable numbering* is a surjective map $\nu : \mathbb{N} \to \mathsf{Comp}(D, \varepsilon)$ for which there exists a total recursive function h with $W_{h(n)} = \{k \in \mathbb{N} \mid \varepsilon_k \sqsubseteq \nu(n)\}$ for all $n \in \mathbb{N}$. ◊

Obviously, for every computable numbering ν of $\mathsf{Comp}(D, \varepsilon)$ there exists a total recursive function t with $\zeta \circ t = \nu$ (namely $t = g \circ h$). On computable numberings one may consider the following preorder

$$\nu_1 \preceq \nu_2 \quad \text{iff} \quad \text{there exists a total recursive } h \text{ with } \nu_2 \circ h = \nu_1.$$

Obviously, a principal numbering is a greatest (w.r.t. $\preceq$) computable numbering from which it is immediate that principal numberings are unique up to recursive reindexing, i.e. whenever ζ and ζ' are principal numberings then $\zeta \circ f = \zeta'$ and $\zeta \circ g = \zeta$ for some total recursive functions f and g.[8] We leave it as an exercise(!) for the inclined reader to verify that for every effectively given domain there does actually exist a principal numbering of its computable elements.

Next we consider a notion of semidecidable subsets for effectively given domains.

Definition 13.16 (extensionally r.e.)
A subset $P \subseteq \mathsf{Comp}(D, \varepsilon)$ is called *extensionally recursively enumerable (e.r.e.)* iff $\{n \in \mathbb{N} \mid \zeta_n \in P\}$ is r.e. where ζ is some principal numbering of $\mathsf{Comp}(D, \varepsilon)$. ◊

Theorem 13.17 (Rice-Shapiro)
For every e.r.e. subset $P \subseteq \mathsf{Comp}(D, \varepsilon)$ it holds that

(1) *for every $d \in P$ there exists an $e \in \mathcal{K}(D) \cap P$ with $e \sqsubseteq d$*
(2) *if $d \in P$ and $d \sqsubseteq d' \in \mathsf{Comp}(D, \varepsilon)$ then $d' \in P$ as well.*

Proof. Let ζ be a principal numbering of $\mathsf{Comp}(D, \varepsilon)$ and f and g be total recursive functions satisfying the requirements (ζ_1) and (ζ_2) of Def. 13.15. *ad* (1) : If d is compact then one may take for e the element d itself. Suppose that d is not compact. For the sake of deriving a contradiction suppose that $e \notin P$ for all compact $e \sqsubseteq d$. As by assumption $d \in \mathsf{Comp}(D, \varepsilon)$ there exists a r.e. set A such that $\varepsilon[A]$ is a chain in D with supremum d. Let h be a total recursive function such that for all $n \in \mathbb{N}$

$$W_{h(n)} = \{m \in A \mid \forall k \leq m.\ \neg T(n, n, k)\}$$

---

[8] *cf.* the *admissible numberings* of partial recursive functions as introduced and studied in [Rogers 1987]

where T is Kleene's T-predicate (see e.g. [Rogers 1987]). Obviously, for every $n \in \mathbb{N}$ the set $\varepsilon[W_{h(n)}]$ is a chain in D and thus its supremum is $\zeta_{g(h(n))}$. Now for all $n \in \mathbb{N}$ we have $\{n\}(n)\uparrow$ iff $\bigsqcup \varepsilon[W_{h(n)}] = d$, i.e.

$$n \notin K \quad \text{iff} \quad \zeta_{g(h(n))} \in P$$

where $K = \{n \in \mathbb{N} \mid \{n\}(n)\downarrow\}$. As P is e.r.e. the complement of K is r.e. in contradiction to the well-known undecidability of the r.e. set K (see [Rogers 1987]).

ad (2) : Suppose $d \in P$ and $d \sqsubseteq d' \in \mathsf{Comp}(D, \varepsilon)$. W.l.o.g. assume that $d \neq d'$. Due to the already established condition (1) there exists an $e \in \mathcal{K}(D) \cap P$ with $e \sqsubseteq d$. Let A be a r.e. set such that $\varepsilon[A]$ is a chain in D with supremum d'. W.l.o.g. assume that $e = \varepsilon_{i_0}$ for some $i_0 \in A$. The partial function

$$\tilde{h}(n, k) = \begin{cases} 0 & \text{if } (n \in K \wedge k \in A) \vee k = i_0 \\ \uparrow & \text{otherwise} \end{cases}$$

is obviously partial recursive. Thus, there exists a total recursive function h with

$$W_{h(n)} = \{k \in \mathbb{N} \mid \tilde{h}(n, k)\downarrow\}$$

for all $n \in \mathbb{N}$. Now if $d' \notin P$ then we have

$$n \notin K \quad \text{iff} \quad W_{h(n)} = \{i_0\} \quad \text{iff} \quad \zeta_{g(h(n))} \in P$$

rendering the complement of K r.e. (since P is .r.e.) in contradiction with the undecidability of the halting problem. $\qquad\square$

Notice that the Rice-Shapiro theorem can be rephrased as follows: e.r.e. subsets of $\mathsf{Comp}(D, \varepsilon)$ are open (w.r.t. the subspace topology on $\mathsf{Comp}(D, \varepsilon)$ induced by the Scott topology on D). It is a straightforward exercise(!) to show that an open subset U of $\mathsf{Comp}(D, \varepsilon)$ is e.r.e. provided $\{n \in \mathbb{N} \mid \varepsilon_n \in U\}$ is r.e.

A fairly immediate consequence of the Rice-Shapiro theorem is the *Myhill-Shepherdson Theorem* whose formulation, however, requires the following notion.

Definition 13.18   (effective morphism of e.g. domains)
A function $f : \mathsf{Comp}(D_1, \varepsilon^{(1)}) \to \mathsf{Comp}(D_2, \varepsilon^{(2)})$ is called an *effective morphism* iff there exists a total recursive function h with $\zeta^{(2)} \circ h = f \circ \zeta^{(1)}$,

i.e. making the diagram

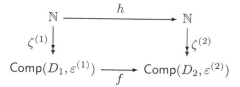

commute, where the $\zeta^{(i)}$ are principal numberings of $\mathsf{Comp}(D_i, \varepsilon^{(i)})$. $\Diamond$

The Myhill-Shepherdson theorem says that every effective morphism of e.g. domains is Scott continuous (i.e. continuous w.r.t. the subspace topologies induced by the Scott topologies).

Theorem 13.19 (Myhill-Shepherdson)
Every effective morphism $f : \mathsf{Comp}(D_1, \varepsilon^{(1)}) \to \mathsf{Comp}(D_2, \varepsilon^{(2)})$ is monotonic and whenever $f(d) \sqsupseteq e_2 \in \mathcal{K}(D_2)$ then there exists an $e_1 \in \mathcal{K}(D_1)$ with $f(e_1) \sqsupseteq e_2$ and $e_1 \sqsubseteq d$.

Proof. Suppose $d \in \mathsf{Comp}(D_1, \varepsilon^{(1)})$ and $e_2 \in \mathcal{K}(D_2)$ with $e_2 \sqsubseteq f(d)$. The set $A_2 = \{n \in \mathbb{N} \mid e_2 \sqsubseteq \zeta_n^{(2)}\}$ is r.e. from which it follows (as f is effective) that the set $A_1 = \{m \in \mathbb{N} \mid e_2 \sqsubseteq f(\zeta_m^{(1)})\}$ is also r.e. Thus, the set $\{d \in \mathsf{Comp}(D_1, \varepsilon^{(1)}) \mid e_2 \sqsubseteq f(d)\}$ is e.r.e. and contains d as an element from which it follows by Theorem 13.17(1) that there exists a compact $e_1 \sqsubseteq d$ with $e_2 \sqsubseteq f(e_1)$.

It remains to show that f is monotonic. For that purpose suppose $d_1, d_2 \in \mathsf{Comp}(D_1, \varepsilon^{(1)})$ with $d_1 \sqsubseteq d_2$. Suppose $e_2 \in \mathcal{K}(D_2)$ with $e_2 \sqsubseteq f(d_1)$. As f is effective the set $\{d \in \mathsf{Comp}(D_1, \varepsilon^{(1)}) \mid e_2 \sqsubseteq f(d)\}$ is e.r.e. and contains d_1 as an element. From Theorem 13.17(2) it follows that $e_2 \sqsubseteq f(d_2)$. As this implication holds for all $e_2 \in \mathcal{K}(D_2)$ we conclude that $f(d_1) \sqsubseteq f(d_2)$ as desired. $\square$

This gives rise to the following characterisation of effective morphims.

Theorem 13.20 *A function $f : \mathsf{Comp}(D_1, \varepsilon^{(1)}) \to \mathsf{Comp}(D_2, \varepsilon^{(2)})$ is an effective morphism if and only if $f = \bar{f} \upharpoonright \mathsf{Comp}(D_1, \varepsilon^{(1)})$ for a (unique) continuous function $\bar{f} : D_1 \to D_2$ computable w.r.t. $\varepsilon^{(1)}$ and $\varepsilon^{(2)}$.*

Proof. An effective morphism $f : \mathsf{Comp}(D_1, \varepsilon^{(1)}) \to \mathsf{Comp}(D_2, \varepsilon^{(2)})$ is Scott continuous by Theorem 13.19. Moreover, the set $\{\langle n, m \rangle \mid \varepsilon_m^{(2)} \sqsubseteq f(\varepsilon_n^{(1)})\}$ is r.e. as there is a recursive function h with $\zeta_{h(n)}^{(1)} = \varepsilon_n^{(1)}$ for all $n \in \mathbb{N}$. Take for $\bar{f}$ the unique continuous function whose (Scott) graph is $\{\langle n, m \rangle \mid \varepsilon_m^{(2)} \sqsubseteq f(\varepsilon_n^{(1)})\}$, i.e. $\bar{f}(d) = \bigsqcup\{\varepsilon_m^{(2)} \mid \varepsilon_n^{(1)} \sqsubseteq d$ and $\varepsilon_m^{(2)} \sqsubseteq f(\varepsilon_n^{(1)})\}$.

For the reverse direction suppose $\bar{f} : D_1 \to D_2$ is continuous and computable w.r.t. $\varepsilon^{(1)}$ and $\varepsilon^{(2)}$. Then the set $A = \{\langle n, m \rangle \mid \varepsilon_m^{(2)} \sqsubseteq \bar{f}(\varepsilon_n^{(1)})\}$ is r.e. Let g be a total recursive function with $W_{g(n)} = \{k \in \mathbb{N} \mid \varepsilon_k^{(1)} \sqsubseteq \zeta_n^{(1)}\}$ and h a total recursive function with $\zeta_{h(n)}^{(2)} = \bigsqcup \varepsilon^{(2)}[W_n]$ whenever $\varepsilon^{(2)}[W_n]$ is directed. Obviously, the set $B = \{\langle n, m \rangle \mid \exists k.\ k \in W_{g(n)} \wedge \langle k, m \rangle \in A\}$ is r.e. as well. Thus, there exists a total recursive function b with $W_{b(n)} = \{m \in \mathbb{N} \mid \langle n, m \rangle \in B\}$. Then for the restriction $f = \bar{f} \upharpoonright \mathsf{Comp}(D_1, \varepsilon^{(1)})$ it holds that $f(\zeta_n^{(1)}) = \zeta_{h(b(n))}^{(2)}$ for all $n \in \mathbb{N}$, i.e.

$$f \circ \zeta^{(1)} = \zeta^{(2)} \circ h \circ b$$

from which it follows that f is an effective morphism since $h \circ b$ is recursive.

As a continuous function on a Scott domain is uniquely determined by its behaviour on compact elements the function $\bar{f}$ is uniquely determined by f since compact elements are in particular computable. $\qquad\square$

Thus, we have established a 1-1-correspondence between computable continuous maps and effective morphisms which generalizes the Myhill-Shepherdson theorem for effective operations and operators as discussed in [Rogers 1987] to effectively given domains.

Moreover, one easily checks (exercise!) that e.r.e. subsets of $\mathsf{Comp}(D, \varepsilon)$ are in 1-1-correspondence with effective morphisms from (D, ε) to the so-called Sierpinski space $\Sigma = \{\bot, \top\}$ w.r.t. the effective presentation $\varepsilon^\Sigma(0) = \bot$ and $\varepsilon^\Sigma(n+1) = \top$. This analogy between topological and recursion-theoretic notions was D. Scott's main motivation for his Domain Theory and lies at the heart of so-called *Synthetic Domain Theory* (SDT) introduced around 1990 independently by Martin Hyland in [Hyland 1990] and Paul Taylor in [Taylor 1991]. SDT aims at axiomatizing *domains as particular sets* within constructive higher order logic or set theory. This slogan of *domains as particular sets* was promoted by D. Scott during the 1980ies and finally taken up by Hyland and Taylor. See [Simpson 2004; Reus and Streicher 1999] for an up-to-date account of SDT.

Bibliography

S. Abramsky, R. Jagadeesan, P. Malacaria *Full Abstraction for* PCF Inform. and Comput. 163, no. 2, 409–470 (2000).

R. Amadio and P.-L. Curien *Domains and λ-Calculi.* Cambridge Univ. Press (1998).

A. Asperti *Stability and computability in coherent domains.* Inform. and Comput. 86, no. 2, 115-139 (1990).

H. Barendregt *The Lambda Calculus. Its Syntax and Semantics.* North Holland (1981).

M. Barr and Ch. Wells *Category theory for computing science.* Prentice Hall International (1990).

G. Berry *Modèles complètement adéquats et stables des lambda calculs typés* Thése d' Etat, Université Paris 7 (1979).

M. H. Escardó PCF *extended with real numbers: a domain-theoretic approach to higher-order exact real number computation.* PhD Thesis, Imperial College, Univ. of London (1997).

E. Griffor, I. Lindström and V. Stoltenberg-Hansen *Mathematical Theory of Domains.* Cambridge Univ. Press (1994).

J. M. E. Hyland *First steps in synthetic domain theory.* in *Category Theory (Como 1990)* pp.131–156, Lecture Notes in Math. 1488 Springer, Berlin, 1991.

M. Hyland, L. Ong *On Full Abstraction for* PCF: *I,II and III.* Inform. and Comput. 163, no. 2, pp.285–408 (2000).

A. Kanda, D. Park *When are two effectively given domains identical?* pp. 563–577, LNCS 67, Springer, Berlin-New York, 1979.

U. de'Liguoro *PCF definability via Kripke logical relations (after O'Hearn and Riecke)* Lab. Inf. ENS, Report LIENS-96-7, Paris (1996).

R. Loader *Finitary PCF is not decidable* TCS 266, no. 1-2, pp. 341-364 (2001).

J. Longley *Sequentially Realizable Functionals* APAL 117, no. 1-3, pp. 1-93 (2002).

S. MacLane *Categories for the Working Mathematician* Second edition. Graduate Texts in Mathematics 5, Springer (1998).

M. Marz *A Fully Abstract Model for Sequential Computation* PhD Thesis, TU

Darmstadt (2000) electronically available from
www.mathematik.tu-darmstadt.de/~streicher/THESES/marz.ps.gz

G. McCusker *Games and full abstraction for a functional metalanguage with recursive types.* CPHC/BCS Distinguished Dissertations, Springer Verlag London (1998).

R. Milner *Fully Abstract Models of Typed λ-Calculi* TCS 4, pp. 1-22 (1977).

D. Normann *On sequential functionals of type 3* to appear in Math. Struct. Comput. Sc. as part of a Festschrift for K. Keimel (2006).

G. Plotkin *LCF considered as a programming language* TCS 5, pp. 223-255 (1977).

G. Plotkin $\mathbb{T}^\omega$ *as a universal domain* J. Comput. System Sci. 17 (1978), no. 2, 209–236.

P. O'Hearn, J. Riecke *Kripke logical relations and PCF* Inf. and Comp. 120, no. 1, pp. 107-116 (1995).

B. Reus, T. Streicher *General synthetic domain theory—a logical approach.* Math. Structures Comput. Sci. 9, no. 2, pp. 177–223 (1999).

J. Riecke, A. Sandholm *A relational account of call-by-value sequentiality* Inf. and Comp. 179, no. 2, pp. 296-331 (2002).

H. Rogers jr. *Theory of recursive functions and effective computability.* 2nd edition, MIT Press, Cambridge, MA, 1987.

A. Rohr *A Universal Realizability Model for Sequential Computation.* PhD Thesis, TU Darmstadt (2002) electronically available from
www.mathematik.tu-darmstadt.de/~streicher/THESES/rohr.ps.gz

Dana S. Scott *A type-theoretical alternative to ISWIM, CUCH, OWHY* unpublished paper from 1969 reprinted in *Böhm Festschrift* TCS 121 No.1-2, pp. 411-440 (1993).

Dana S. Scott *Relating theories of the λ-calculus* in *To H. B. Curry: essays on combinatory logic, lambda calculus and formalism* pp. 403-450 Academic Press, London-New York (1980).

K. Sieber *Reasoning about sequential functions via logical relations* in *Applications of category theory in computer science* pp.258–269 London Math. Soc. Lecture Note Ser., 177, CUP (1992).

A. Simpson *Computational Adequacy for Recursive Types in Models of Intuitionistic Set Theory.* Ann. Pure Appl. Logic 130, no. 1-3, pp. 207–275 (2004).

A. Stoughton *Equationally fully abstract models of PCF* Proc. of *Fifth International Conference on the Mathematical Foundations of Computer Science* Springer Lecture Notes in Comput. Sci. 442, pp.271–283 (1990)

T. Streicher *A universality theorem for PCF with recursive types, parallel-or and $\exists$.* Math. Struct. Comput. Sc. 4, no. 1, pp.111-115 (1994).

T. Streicher, B. Reus *Classical logic, continuation semantics and abstract machines.* J. Funct. Programming 8 (1998), no. 6, 543–572.

P. Taylor *The fixed point property in synthetic domain theory.* in *Proc. 6th IEEE Symposium on Logic in Computer Science*, pp.152-160 (1991).

G. Winskel *The formal semantics of programming languages. An introduction.* MIT Press, Cambridge MA (1993).

Index